Springer Complexity

Springer Complexity is an interdisciplinary program publishing the best research and academic-level teaching on both fundamental and applied aspects of complex systems—cutting across all traditional disciplines of the natural and life sciences, engineering, economics, medicine, neuroscience, social and computer science.

Complex Systems are systems that comprise many interacting parts with the ability to generate a new quality of macroscopic collective behavior the manifestations of which are the spontaneous formation of distinctive temporal, spatial or functional structures. Models of such systems can be successfully mapped onto quite diverse "real-life" situations like the climate, the coherent emission of light from lasers, chemical reaction–diffusion systems, biological cellular networks, the dynamics of stock markets and of the internet, earthquake statistics and prediction, freeway traffic, the human brain, or the formation of opinions in social systems, to name just some of the popular applications.

Although their scope and methodologies overlap somewhat, one can distinguish the following main concepts and tools: self-organization, nonlinear dynamics, synergetics, turbulence, dynamical systems, catastrophes, instabilities, stochastic processes, chaos, graphs and networks, cellular automata, adaptive systems, genetic algorithms and computational intelligence.

The three major book publication platforms of the Springer Complexity program are the monograph series "Understanding Complex Systems" focusing on the various applications of complexity, the "Springer Series in Synergetics", which is devoted to the quantitative theoretical and methodological foundations, and the "SpringerBriefs in Complexity" which are concise and topical working reports, case-studies, surveys, essays and lecture notes of relevance to the field. In addition to the books in these two core series, the program also incorporates individual titles ranging from textbooks to major reference works.

Editorial and Programme Advisory Board

Understanding Complex Systems

Founding Editor: Scott Kelso

Future scientific and technological developments in many fields will necessarily depend upon coming to grips with complex systems. Such systems are complex in both their composition—typically many different kinds of components interacting simultaneously and nonlinearly with each other and their environments on multiple levels—and in the rich diversity of behavior of which they are capable.

The Springer Series in Understanding Complex Systems series (UCS) promotes new strategies and paradigms for understanding and realizing applications of complex systems research in a wide variety of fields and endeavors. UCS is explicitly transdisciplinary. It has three main goals: First, to elaborate the concepts, methods and tools of complex systems at all levels of description and in all scientific fields, especially newly emerging areas within the life, social, behavioral, economic, neuro- and cognitive sciences (and derivatives thereof); second, to encourage novel applications of these ideas in various fields of engineering and computation such as robotics, nano-technology and informatics; third, to provide a single forum within which commonalities and differences in the workings of complex systems may be discerned, hence leading to deeper insight and understanding.

UCS will publish monographs, lecture notes and selected edited contributions aimed at communicating new findings to a large multidisciplinary audience.

For further volumes:
http://www.springer.com/series/5394

Christian Walloth · Jens Martin Gurr
J. Alexander Schmidt
Editors

Understanding Complex Urban Systems: Multidisciplinary Approaches to Modeling

Springer

Editors
Christian Walloth
Institute of City Planning and Urban Design
University of Duisburg-Essen
Essen
Germany

J. Alexander Schmidt
Institute of City Planning and Urban Design,
Joint Center "Urban Systems"
University of Duisburg-Essen
Essen
Germany

Jens Martin Gurr
Department of Anglophone Studies,
Joint Center "Urban Systems"
University of Duisburg-Essen
Essen
Germany

ISSN 1860-0832 ISSN 1860-0840 (electronic)
ISBN 978-3-319-02995-5 ISBN 978-3-319-02996-2 (eBook)
DOI 10.1007/978-3-319-02996-2
Springer Cham Heidelberg New York Dordrecht London

Library of Congress Control Number: 2013954026

Printed on acid-free paper

Springer is part of Springer Science+Business Media (www.springer.com)

Preface

Understanding Complex Urban Systems is targeted at researchers and students of urban and regional studies. This volume is also targeted at urban strategists and urban planning professionals who have experienced the limits of traditional planning approaches and are now seeking to understand the city as a complex system. The reader is presumed to possess a basic knowledge of complex systems approaches. Although no specific knowledge is required, some insight into one or more complex systems concepts such as, e.g., agent-based modeling, the Viable Systems Model, or organized and disorganized complexity may facilitate the reader's understanding of the papers presented in this volume.

The present volume aims to advance the understanding of cities as complex systems and to support urbanists and urban developers in better understanding characteristics of urban complexity and in dealing with this complexity using appropriate methods. The papers of this volume show how systems and complexity theories can be applied in urban research and urban development. A further aim of this book is to advance the discourse concerning current and in-principle limits of modeling and planning in complex systems by presenting a wide variety of often complementary approaches from evolutionary economics to literary studies, to systemic urban development, and further to agent-based and system dynamics modeling as well as stochastic modeling. The editors and contributors share the conviction that the complexity of the city as an object of studies makes complementary approaches from various disciplines essential to an adequate understanding.

Thus, the focus of this volume lies on different modeling approaches originating in various disciplines. The articles and essays presented provide some insights from the forefront of complex urban systems research, e.g., modeling of human behavior during disaster situations, dealing with unpredictable new qualities, and handling of uncertain future events. The volume thus seeks both to advance theoretical and methodological thinking in research on complex urban systems and to suggest ways of dealing with complexity in practice.

Essen, Bonn, Brussels
August 2013

Christian Walloth
Jens Martin Gurr
J. Alexander Schmidt

Acknowledgments

The editors would like to thank Stefan Blachfellner and Wolfgang Hofkirchner for the organization of the EMCSR—European Meeting of Cybernetics and Systems Research in Vienna in 2012, which provided us with the occasion for the first symposium on Urban Systems Research, out of which the idea for this book was born. We also very gratefully acknowledge the help of Judit Knisz, Florin Leonte and Stefanie Caeners in reviewing and editing this volume.

Further thanks are due to our publisher, Springer, for a smooth and hassle-free production process.

We are most grateful, however, to the authors for their contributions, for the smooth and pleasant cooperation in the editing process, and for their unusual willingness to respond to queries and requests.

Finally, we would like to thank our families for their understanding and support during the time of editing and finalizing this book.

Contents

Contributors

Funda Atun Department of Architecture and Urban Studies, Politecnico di Milano, Via Bonardi 3, 20133 Milan, MI, Italy, e-mail: funda.atun@polimi.it

Ernst Gebetsroither-Geringer Energy Department, Austrian Institute of Technology, Giefinggasse 2, 1210 Vienna, Austria, e-mail: Ernst.Gebetsroither@ait.ac.at

Jens Martin Gurr Department of Anglophone Studies, Joint Center "Urban Systems", University of Duisburg-Essen, Universitätsstraße 12, 45141 Essen, Germany, e-mail: jens.gurr@uni-due.de

Tom Kauko Department of Geography, Norwegian University of Science and Technology—NTNU, 7491 Trondheim, Norway, e-mail: tom.kauko@svt.ntnu.no

Xosé L. Martínez Suárez University of A Coruña, E.T.S. Arquitectura, Campus A Zapateira, 15071 A Coruña, Spain, e-mail: xlms28@yahoo.es

José Pérez Ríos University of Valladolid, E.T.S.I. Informatica, Campus Miguel Delibes, 47011 Valladolid, Spain, e-mail: rios@uva.es

Rüdiger Schultz Department of Mathematics, University of Duisburg-Essen, Forsthausweg 2, 47057 Duisburg, Germany, e-mail: ruediger.schultz@uni-due.de

Christian Walloth Institute of City Planning and Urban Design, University of Duisburg-Essen, Universitätsstraße 15, V15 S04 D03, 45141 Essen, Germany, e-mail: christian.walloth@stud.uni-due.de

Hans-Werner Wehling Institute of Geography, University of Duisburg-Essen, Schützenbahn 70, SR 107, 45127 Essen, Germany, e-mail: hans-werner.wehling@uni-due.de

Introduction: Towards a Transdisciplinary Understanding of Complex Urban Systems

Jens Martin Gurr and Christian Walloth

Abstract While complex systems research has rapidly advanced over the past decades, the planning disciplines are still at an early stage of devising new methods and instruments informed by complexity science. In order to see cities in a new way, to recognize ongoing processes, and to understand effects of interventions, the practice of planning and urban development is in need of transdisciplinary approaches to observing and understanding complex urban systems. It is time now, we believe, to review the various approaches offered by different disciplines and to start discussing in what way they can be regarded as complementary—or can even be integrated—in order to better understand complex urban systems.

Keywords Complex systems research · Urban systems · Urban planning and development · System modeling · Transdisciplinary approaches

1 Understanding Cities as Complex Systems: Current Challenges and the Role of Urban Systems Research

Global urbanization with the numerous challenges it entails requires the joint efforts of experts from an array of disciplines to help develop new ways of dealing with the city—and given the multiple interdependencies between demographic,

---

J. M. Gurr (✉)
Department of Anglophone Studies, Joint Center "Urban Systems", University of Duisburg-Essen, Universitätsstraße 12, 45141 Essen, Germany
e-mail: jens.gurr@uni-due.de

C. Walloth
Institute of City Planning and Urban Design, University of Duisburg-Essen, Universitätsstraße 15, V15 S04 D03, 45141 Essen, Germany
e-mail: christian_walloth@complexurbansystems.net

C. Walloth et al. (eds.), *Understanding Complex Urban Systems: Multidisciplinary Approaches to Modeling*, Understanding Complex Systems,
DOI: 10.1007/978-3-319-02996-2_1,

climatic, infrastructural, economic, ecological, cultural, legal, social and innumerable further developments and requirements in urban systems, research on complex urban systems is called upon to contribute to the understanding and the management of urban complexity. Complex urban systems may thus be the ideal (and arguably most pressing) field in which to explore the implications of the revolution brought about by systems science, complexity science and non-linear dynamics since the last decades of the twentieth century.

While the planning disciplines are still at an early stage of devising new methods and instruments in engaging with complex urban systems, complex systems research has rapidly advanced over the past decades (cf. Sect. 2.1 of this Introduction) and the application of complex systems theory in the field of urban research has recently gained momentum. For instance, Haase et al. (2011, p. 82) recognize the "simultaneity and interrelatedness of a number of processes" and the "complexity of … development" in post-socialist cities. Deppisch and Schaerffer (2011, pp. 25–26) discuss the "characteristics of complex cities", mentioning, inter alia, "non-linearity and emergence" in the context of urban resilience. In another contribution on "resilience and transformation", Prosperi and Morgado (2011, pp. 820 ff.) dedicate a large section of their paper to "key primitive concepts of complexity theory".

One of the most ambitious recent attempts to provide interdisciplinary urban research with an integrating paradigm, urban sociologist Frank Eckardt's *Die komplexe Stadt: Orientierungen im urbanen Labyrinth*, even suggests that approaches from complexity theory might serve as the central paradigm around which interdisciplinary urban studies might be organized (cf. Eckardt 2009). In a similar vein, Portugali argues that complexity theories have the potential to bridge the gap between quantitative 'hard' sciences on the one hand and the 'soft' sciences and approaches in the humanities on the other hand (Portugali 2006).

In seeking to approach the city from a complex systems perspective, the understanding and application of systems phenomena such as non-linear, feedback-driven developments, self-organization, emergence of new qualities, qualitative behaviors, the autopoietic drive to self-sustainment, and phase transitions into new systems are crucial. Take, for instance, the case of phase transitions: Such phase transitions are marked by significant uncertainty and indeterminacy regarding further developments. Currently developed scenarios for the ecological future of cities, for instance, simulate different desirable as well as undesirable urban futures, such as "business as usual" or the "sustainable and climate-friendly city", the "city of citizens" or the "eco-dictatorship". This shows the uncertainty and imponderability as well as the developments planning takes into account (and those it does not), possible directions of urban development and how uncertain any vision of urban futures currently is. What we know from the natural sciences is that phase transitions are particularly sensitive to even small new impulses from outside; phases in which the wrong kind of interventions can have catastrophic consequences. Seen more optimistically, bearing in mind the notion of *kairos*, phase transitions are intervention-friendly and open up opportunities for triggering developments—at the right moment, in the right place, with the right kind of (limited) effort (cf. Schmidt 1990).

It seems clear that neither a quantitative understanding nor the qualitative and descriptive approaches in some branches of the social sciences and in the humanities alone will suffice in understanding and dealing with complex urban systems: What is needed is a transdisciplinary approach to observing complex urban systems and to seeing them in a new way, to recognizing ongoing processes, to understanding impulses and their effects, but above all, to thinking the city anew. The practice of planning and urban development, as well as applied research, cannot passively stand by in all this and merely observe. It needs to engage in dialog with research on complex urban systems and, based on a better understanding of urban complexity, it should actively intervene. The understanding of the city as a complex system, above all, calls for planning to be critical towards rapidly dispensed cookie-cutter recipes and 'best practice' models (because different systems respond differently to similar interventions).

Understanding Complex Urban Systems thus takes as its point of departure the insight that the complexity of urban systems cannot possibly be understood—let alone 'managed'—by sectoral and disciplinary approaches alone. In the kind of joint research effort we envisage, the central questions are: How can the various disciplines contribute to inter- and transdisciplinary research on complex urban systems and what are the implications for the practice of urban planning? What strategies are being used in dealing with complexity? How can it be 'modeled'—represented, conceptualized, managed, or reduced?

2 Multidisciplinary Approaches to Urban Modeling: Status Quo, the Present Volume—and Beyond

This book attempts to contribute to current discussions on the modeling of complex urban systems. Starting out from systems and complexity research that has evolved since the mid-twentieth century, complex systems have been approached from various disciplinary angles more or less independently and developed around different aspects, e.g., general systems phenomena, chaotic behavior, fractal structures, or systemic interventions. It is time now, we believe, to review the various approaches offered by different disciplines and to start discussing in what way they can be regarded as complementary, or can even be integrated. In this vein, the present volume brings together a variety of approaches to understanding and modeling complex urban systems. To be sure, however, this book, featuring contributions still strongly rooted in disciplinary work, is a first step only.

2.1 Status Quo

The system and complexity sciences, and along with them approaches to modeling complex systems, have shown tremendous progress since their beginnings around the mid-twentieth century. Von Bertalanffy's (1969) *General Systems Theory*, a collection of papers authored between 1940 and 1969, is still a great compendium on the phenomenology of complex systems. The more formalized approaches of the same era have led to the first system dynamics models, e.g., a model on 'Urban Dynamics' by Forrester (1969) and the World 3 model by Meadows (1973) that led to the publication of the famous *The Limits to Growth* (Meadows et al. 1972).

From the 1970s onwards, 'soft' system (or 'systemic') approaches, seeing systems as ultimately constructed by the observer, gained ground. For instance, Vester's Biocybernetic Approach, applied to modeling the metropolitan region of Frankfurt am Main (Vester 1976, Vester and von Hesler 1980), yields 'soft' system models. A decade ago, Hummelbrunner et al. (2002) comprehensively compiled the variety of systemic approaches applicable to urban systems.

Complexity sciences continued down the 'hard sciences' road, based on formalized, mathematical methods. What is central here, too, are the insights into complexity from non-linear dynamics (a.k.a. 'chaos research'), with notions such as the 'Butterfly Effect'—a small change here and now causing a large effect elsewhere and at a later moment, potentially with catastrophic consequences—or the study of fractal forms—some urban morphologies have strong resemblances to fractals (cf., e.g., Batty and Longley 1994). Somewhat different approaches were pursued by Nicolis and Prigogine (1977) in their theory of dissipative systems, i.e. systems that self-organize into a higher order if fed with energy, and by Haken's (1977) theory of synergetics that focuses on order parameters, i.e. top-down 'enslaving' rules, as a key systems phenomenon.

However, since Forrester's Urban Dynamics applied to Boston, Vester's Biocybernetic Approach applied to Frankfurt am Main, and Allen's application of dissipative systems theory (e.g., Allen and Sanglier 1981) and Weidlich's application of synergetics (Weidlich 1987) to population migration yielding settlement patterns reminiscent of central places (Christaller 1933), there seems to have been a large gap with little application of complexity and systems theories to modeling cities until fairly recently. Over the past decade or so, it is true, there has again been significant progress in broadening and refining the methodologies for and approaches to the quantitative modeling of complex urban systems (cf., e.g., Bretschneider and Kimms 2011; Drechsel and Kimms 2008; Handschin et al. 2006; Kimms and Maassen 2011; Kuhn and Schultz 2009, Schultz and Wagner 2009), in deepening the theoretical understanding of cities as complex systems (cf., e.g., Albeverio et al. 2008; Batty 2009; Portugali 2008, 2011; Portugali et al. 2012), or in illuminating the implications for urban planning and urban development (cf., e.g., de Roo et al. 2012; de Roo and Silva 2010, Schmidt and Walloth 2012); moreover, there are a number of studies on the modeling of urban complexity in literature and cultural production (cf., e.g., Gurr 2011a, b; Gurr and

Raussert 2011). Finally, long after the (indeed very basic) World 3 model, there are again attempts to understand underlying patters and to build comprehensive world models based on large amounts of quantitative data (e.g., Bettencourt et al. 2007; FuturICT 2013). However, brute force—processing more and more data and running statistical analyses and models—might not be the ultimate way forward. After all, Weaver (1948) already distinguished between systems that can be handled statistically—systems of 'disorganized complexity'—and systems that cannot be handled statistically—systems of 'organized complexity'. To be sure, cities belong into the realm of organized complexity, as was pointed out by Jacobs (1961, pp. 428 ff.) based on Weaver's earlier contribution (see also Wehling 2014, in this volume).

Moreover, there is still a lack of well-founded conceptual thinking on the methodological foundations and the strategies of modeling urban complexity across the disciplines. Cooperation between the primarily quantitative disciplines on the one hand and more qualitative approaches from the humanities and some branches of the social sciences is hardly taking place—although it is frequently called for and claimed to be necessary both in research (cf., e.g., Eckardt 2009; Portugali 2012) as well as political research agendas (not least on the EU level).

It is primarily to facilitate such research that the University of Duisburg-Essen in 2008 established the Joint Center Urban Systems as a centrally designated priority research area. Here, some 70 researchers from all eleven faculties of the university—from medicine and the natural sciences via urban planning and engineering, economics and the social sciences to the humanities and educational sciences—engage in interdisciplinary urban research. In addition to a broad range of issues relevant to cities and metropolitan regions worldwide, it is especially the understanding of urban complexities that is a shared concern in much research in the context of this Joint Center.

We believe that in order to understand and model complex urban systems, researchers across the disciplines have to discuss a number of key questions such as the following:

- How is 'urban complexity' understood, defined and 'modeled' in various disciplines? Which research questions are derived from these definitions and conceptualizations in the various disciplines?
- What is the logical status of the model in the respective discipline? Is it the result of scientific endeavors, as in an attempt mathematically to model an energy system, or is a 'model' developed elsewhere the object of study, as in the study of cultural productions as 'models' of urban complexity in literary studies or of visual models of social cartography as studied, for instance, by a social historian?
- Which sub-systems or combinations of sub-systems are being modeled by researchers of various disciplines—e.g., is the city considered as a social, ecological, economic, technical, or cultural system?

Besides, we maintain that practitioners need to discuss related questions, such as, for instance, who decides which parameters to include into decision-making processes—and which to neglect.

2.2 The Present Volume

One of the various initiatives developed out of the Joint Center led to the organization of the symposium Urban Systems Research, which first took place at the University of Vienna in the context of the 2012 EMCSR—European Meeting of Cybernetics and Systems Research. The aim of this bi-annual symposium is to bring together international urban systems and complexity researchers. This volume is based on selected papers from that symposium, centered around key issues in modeling complex urban systems, plus two additional articles specially commissioned for this project.

In what follows, experts from the fields of urban and spatial planning, ecology, urban geography, real estate analysis, organizational cybernetics, stochastic optimization, and literary studies, as well as specialists in various systems approaches and in transdisciplinary methodologies of urban analysis address a number of questions we consider central to advancing the discussion on "Multidisciplinary Approaches to Urban Modeling":

1. How can complex urban systems be described and modeled and what is the relation to urban reality these descriptions and models claim? Do they reduce phenomena to idealized essentials to point out selected system interdependencies or do they seek fully to model a system?
2. What are the—normative or descriptive—aims and interests pursued with different types of models? Are they designed fundamentally to understand interdependencies within or explain the functioning of urban systems—or do they seek to aid decision making in concrete situations?
3. How can central non-quantifiable phenomena that are especially relevant to a humanities understanding of urban complexity be integrated into the dominant types of models? To what extent can they complement mathematical models?

In keeping with the aim of the collection to foster transdisciplinary research on complex urban systems, all contributors were specifically requested to write contributions not only addressing specialists in their respective disciplines. Thus, while engaging with the state-of-the-art of their respective fields, the contributions are specifically written for both experts from a broad range of disciplines as well as for urban practitioners who feel the need for new approaches given the uncertainty of current developments.

The first contributions discuss various methodological concerns in the modeling of complex urban systems: In his article on "Multimethod Modeling and Simulation Supporting Urban Planning Decisions", **Ernst Gebetsroither-Geringer** discusses reasons for the increasing relevance of multimethod simulation and illustrates the main advantages and disadvantages of combining different modeling methods. The article introduces the MASGISmo simulation platform as an example of specialized software tools that support multimethod modeling. The theoretical discussion is supported with results obtained from different simulation projects. The paper shows how an urban development model using a multimethod approach can support policy makers and urban planners in implementing viable planning measures in complex systems. Agent-based modeling (ABM) is shown to enable model users to interpret and react to information from different levels of the urban spatial hierarchy within the simulation. The value added by combining ABM with System Dynamics (SD) modeling along with the use of GIS data is also shown. Finally, the paper discusses how a new way of real stakeholder interactivity within the simulation can be achieved in order to improve the model.

From a mathematical perspective, **Rüdiger Schultz** in "Uncertainty in Urban Systems: How to Optimize Decision Making Using Stochastic Programming" presents stochastic optimization as a methodology for mathematically rigorous and computationally efficient treatment of optimization problems in which data are uncertain but may have known probability distributions. Since optimization problems in urban systems often contain uncertain data, stochastic optimization can be an extremely valuable approach in reducing the complexity of urban systems; conversely, urban systems research provides challenging and practical sample problems for stochastic optimization and thus lends itself as a field of application. The article gives an introduction to stochastic programming resorting to reduced mathematical formalism and real-life application examples from the operation of urban infrastructure and utilities.

In "Understanding Effects of Complexity in Cities During Disasters", **Funda Atun** takes as her point of departure an understanding of cities as complex systems made up of highly heterogeneous and interconnected sub-systems (e.g., infrastructure, economy, health system, etc.), including autonomous entities (individuals) engaged in innumerable non-linear interactions. When such a complex system is struck by a hazard, interrupted interconnections between sub-systems prevent it from functioning normally. But while vulnerability analysis can help to assess which sub-systems and/or interconnections are vulnerable to disasters and may create disruption to the urban system, it is impossible to foresee exactly how a disruption in a sub-system might affect the urban system. In addition to the interruption among physical elements, the reactions of individuals—whether, for instance, they receive and understand the necessary information, follow orders, and properly support an evacuation process—are not entirely predictable. In order to further the understanding of cities during disasters, this contribution gives examples of consequences of uncertain interactions between urban systems and unforeseen reactions of humans during disasters such as the 1995 Kobe earthquake, the 2005 Rita and Katrina Hurricanes, and the 2010 eruption of Icelandic volcano Eyjafjallajokull.

In the final methodological essay on the predominantly quantitative side, **Tom Kauko** in a contribution entitled "Towards Evolutionary Economic Analysis of Sustainable Urban Real Estate: Concept of a Research Strategy, Exemplified on

House Price Modeling Using the Self-organizing Map, Interviews and Field Inspection" outlines an evolutionary economics approach to understanding real estate developments. He takes his cue from the observation that evolutionary economics has begun to complement and partly to replace neoclassical economics as the most widely accepted framework for economic modeling. Nonlinear and iterative, evolutionary frameworks take as their starting point the divergence of routines and outcomes of business activity over time and assume that this diversification subsequently leads to a selection-of-the-fittest mechanism. In the case of real estate analysis, a feasible method of predicting market outcome and the behavior of actors such as developers and investors needs to apply both real estate transaction price data and expert interviews about actor motivations and tendencies. On top of the price development target, urban sustainability is incorporated as a more qualitative target, as the aim is to compare the price development with various sustainability aspects across different locations and typical market segments. This demonstration is based on a modeling approach known as the self-organizing map (SOM).

The second group of essays provide regional or local case studies: **Hans-Werner Wehling's** discussion of "Organized and Disorganized Complexities and Socio-Economic Implications in the Northern Ruhr Area" provides a regional case study: The present-day Ruhr region is characterized by a general south-north (regional) divide in terms of urban and social structures. The contribution argues that the multi-phase processes of nineteenth and early twentieth-century industrialization and of post-war de-industrialization, the impact of the interests and decisions of influential groups in various alliances, as well as various phases of immigration have been disrupting the urban system on different scales. As a result, different urban systems of (dis)organized complexity came into being within the urban system of the region as a whole. Finally, the essay shows how current planning approaches try to do justice to these forms of urban complexity.

On a local scale, **Xosé L. Martínez Suárez** and **José Pérez Ríos** in their contribution entitled "An Organizational Cybernetics Approach to University Planning in an Urban Context: Four Intervention Experiences" approach urban interventions from an Organizational Cybernetics (OC) perspective. They take their cue from the fact that urban planning interventions may be influenced by extremely diverse fields and phenomena—demographic, legislative, economic, ecological, sociological, architectural, urbanistic, academic, and many more. Planning thus needs to do justice to the complexity generated by the potential interrelations among these domains. The essay presents an example of how a public university, the University of A Coruña (UDC) in Galicia, Spain, faced this complexity by means of Organizational Cybernetics. The authors show how the use of OC helped in the identification of various needs to be considered in a planning-related intervention in a university campus and in influencing the project design so as to allow for an increase in the variety of services responding to these needs. After a short description of some OC concepts and of the systemic methodological framework that has been used in this work, the report presents four urbanistic and architectonic interventions on the UDC campuses as case studies in the application of this framework.

The final two contributions engage with the limits of current planning and (quantitative) modeling: **Christian Walloth** in "Emergence in Complex Urban Systems—Blessing or Curse of Planning Efforts?" addresses the problem that current urban planning practices largely rely on the assumption that identical interventions will create identical results in comparable cities. This simplistic assumption is highly problematic in complex urban systems where emergent, i.e. unplanned and ultimately unpredictable qualities lead to successive qualitative changes of the environment: Where emergent qualities cannot be foreseen or planned, and where the outcome of urban interventions depends on emergent qualities, urban planning practice has to find ways of taking emergent qualities into account. The contribution discusses the potentially significant impacts of emergent qualities using examples from non-capital cities in Central and Eastern Europe. The author calls for an increase in sensitivity for detecting emergent qualities and concludes that emergence of new qualities can be a source of a city's unique characteristics, setting it apart from other cities merely trying to re-implement 'cookie cutter' approaches that may have worked elsewhere.

The concluding contribution, **Jens Martin Gurr's** essay "'Urban Complexity' from a Literary and Cultural Studies Perspective: Key Cultural Dimensions and the Challenges of 'Modeling'" continues the discussion on the limitations of current modeling and planning strategies by sketching a literary and cultural studies approach to urban complexity. It briefly comments on the neglect of cultural phenomena in much urban modeling, discusses key characteristics of urban complexity from a literary and cultural studies perspective and relates these to technical and mathematical notions of complexity. After engaging with the need for reduction and compression in modeling and with the resulting limitations of urban models, a more specifically 'literary' section discusses simultaneity as a central characteristic of urban complexity and a challenge to narrative representation and points out a number of literary strategies in the 'modeling' of urban complexity in literature. The article sets out to show what literary studies can contribute to truly interdisciplinary work on urban complexity and urban systems, arguing that literary studies specifically bring back into the discussion those features of urban complexity that resist modeling but that are nonetheless crucial to a differentiated understanding of the functioning of urban systems.

2.3 Outlook

This volume, to be sure, is but a first step in our inter- and transdisciplinary endeavors to understand complex urban systems. While the contributions in the present volume gesture towards transdisciplinary approaches and indicate perspectives for further research, we hope to build on this work in future volumes on various aspects of complex urban systems research. In order to advance the discussion about various approaches to modeling complex urban systems, future work might engage with questions such as the following:

- Which issues and strategies in dealing with urban complexity are already cross-disciplinary? What are the interfaces between disciplines and what bridges can be built here?
- Which of the issues and strategies specific to the individual disciplines dealing with urban complexity can be regarded as complementary? What are the implications of the specific sign systems in which these models are mediated? Are there parallels or analogies between quantitative and narrative models?
- According to which criteria do we make decisions concerning parameters to be included or excluded? To what extent are selection criteria for inclusion and exclusion complementary in different disciplines?
- How can the insights from various types of models be integrated?

One possible suggestion that can be made at this point is that it might be worth studying in more detail to what extent quantitative, visual and narrative models of complex urban systems are comparable and to what extent they might be harmonized or integrated. Going beyond an analysis of the achievements and shortcomings of various conceptualizations of urban modeling and the problem-solving strategies derived from them, we believe it would be worth exploring an integrated conceptualization. Our current hypothesis is that these various types of models are not to be seamlessly integrated into one overarching model—for instance, the modeling of complex urban systems in literary texts, aiming to capture that which is individual, specific, historically and locally unique, already in its underlying interest and thus in the selection decisions is more or less directly opposed to the approach in quantitative modeling seeking to abstract and generalize. Rather, we hypothesize, these different approaches to modeling are to be seen as complementary ways of conceptualizing urban reality, which lend themselves to illuminating very diverse aspects of urban complexity. Such different models might rather complement each other, for instance by mutually setting off shortcomings and by filling in each other's blind spots. This kind of integration of various approaches to modeling, we believe, will be crucial to a more refined understanding of complex urban systems. It is our hope that the present volume may help to further this type of research.

References

Albeverio, S., Andrei, D., Giordano, P., Cancheri, A. (eds.): The Dynamics of Complex Urban Systems: An Interdisciplinary Approach. Physica, Heidelberg (2008)

Allen, P.M., Sanglier, M.: Urban evolution, self-organization, and decisionmaking. Environ. Plan. A **13**, 167–183 (1981)

Batty, M.: Cities as complex systems: scaling, interaction, networks, dynamics and urban morphologies. In: Meyers, R.A. (ed.) Encyclopedia of Complexity and Systems Science, vol. 1, pp. 1041–1071. Springer, Berlin (2009)

Batty, M., Longley, P.: Fractal Cities: A Geometry of Form and Function. Academic Press, London (1994)
von Bertalanffy, L.: General System Theory: Foundations, Development, Applications. George Braziller, New York (1969)
Bettencourt, L.M.A., Lobo, J., Helbing, D., Kühnert, C., West, G.B.: Growth, innovation, scaling, and the pace of life in cities. PNAS **107**(17), 7301–7306 (2007)
Bretschneider, S., Kimms, A.: A basic mathematical model for evacuation problems in urban areas. Transp. Res. Part A: Policy. Pract. **45**(6), 523–539 (2011)
Christaller, W.: Die zentralen Orte in Süddeutschland. Gustav Fischer, Jena (1933)
Deppisch, S., Schaerffer, M.: Given the complexity of large cities, can urban resilience be attained at all? In: Müller, B. (ed): German Annual of Spatial Research and Policy 2010. Urban Regional Resilience: How Do Cities and Regions Deal with Change? pp. 25–34. Springer, Berlin (2011)
Drechsel, J., Kimms, A.: Task force deployment for big events. Saf. Sci. **46**(9), 1289–1305 (2008)
Eckardt, F.: Die komplexe Stadt: Orientierungen im urbanen Labyrinth. Verlag für Sozialwissenschaften, Wiesbaden (2009)
Forrester, J.W.: Urban Dynamics. MIT Press, Boston (1969)
FuturICT (2013): http://www.futurict.eu/
Gurr, J.M.: The literary representation of urban complexity and the problem of simultaneity: a sketchy inventory of strategies. In: Gurr, J.M., Raussert, W. (eds.) Cityscapes in the Americas and Beyond: Representations of Urban Complexity in Literature and Film, pp. 11–36. Trier and Bilingual Press, Tempe, AZ, WVT (2011)
Gurr, J.M.: The politics of representation in hypertext docufiction: multi-ethnic Los Angeles as an emblem of 'America' in Norman M. Klein's Bleeding Through: Layers of Los Angeles 1920–1986. In: Raab, J., Thies, S., Noll-Opitz, D. (eds.) Screening the Americas: Narration of Nation in Documentary Film/Proyectando las Américas: Narración de la nación en el cine documental. pp. 153–171. WVT, Trier and Bilingual Press, Tempe, AZ (2011b)
Gurr, J.M., Raussert, W. (eds.): Cityscapes in the Americas and Beyond: Representations of Urban Complexity in Literature and Film. Trier and Bilingual Press, Tempe, AZ, WVT (2011)
Haase, A., Vaishar, A., Węcławowoicz, G.: The post-socialist condition and beyond: framing and explaining urban change in East Central Europe. In: Haase, A., Steinführer, A., Kabisch, S., Grossmannn, K., Hall, R.(eds.): Residential Change and Demographic Challenge: The Inner City of East Central Europe in the 21st Century. pp. 63–83. Ashgate, Burlington (2011)
Haken, H.: Synergetics, An Introduction, Nonequilibrium Phase Transitions and Self-Organization in Physics, Chemistry and Biology. Springer, Berlin (1977)
Handschin, E., Neise, F., Neumann, H., Schultz, R.: Optimal operation of dispersed generation under uncertainty using mathematical programming. Int. J. Electr. Power Energy Syst. **28**, 618–626 (2006)
Hummelbrunner, R., Lukesch, R., Baumfeld, L.: Systemische Instrumente für die Regionalentwicklung. ÖAR-Regionalberatung, Graz (2002)
Jacobs, J.: The Life and Death of Great American Cities. Random House, New York (1961)
Kimms, A., Maassen, K.: Optimization and simulation of traffic flows in the case of evacuating urban areas. OR Spectrum **33**, 571–593 (2011)
Kuhn, S., Schultz, R.: Risk neutral and risk averse power optimization in electricity networks with dispersed generation. Math. Methods Oper. Res. **69**, 353–367 (2009)
Meadows, D.H.: Club of Rome: The Limits to Growth: A Report for the Club of Rome's Project on the Predicament of Mankind. Universe, New York (1972)
Meadows, D.L.: The Dynamics of Growth in a Finite World: A Technical Report on the Global Simulation Model World 3. Thayer School of Engineering, Darmouth (1973)
Nicolis, G., Prigogine, I.: Self-Organization in Nonequilibrium Systems: From Dissipative Structures to Order through Fluctuations. Wiley, New York (1977)
Portugali, J.: Complexity theory as a link between space and place. Environ.Plan. A **38**, 647–664 (2006)

Portugali, J.: A structural-cognitive approach to urban simulation models. In: Albeverio, S., Andrei, D., Giordano, P., Cancheri, A. (eds.) The Dynamics of Complex Urban Systems: An Interdisciplinary Approach, pp. 357–372. Physica, Heidelberg (2008)
Portugali, J.: Complexity, Cognition and the City. Springer, Berlin (2011)
Portugali, J.: Complexity theories of cities: achievements, criticism and potentials. In: Portugali, J., Meyer, H., Stolk, E., Tan, E. (eds.) Complexity Theories of Cities Have Come of Age: An Overview with Implications to Urban Planning and Design, pp. 47–62. Springer, Berlin (2012)
Portugali, J., Meyer, H., Stolk, E., Tan, E. (eds.): Complexity Theories of Cities Have Come of Age: An Overview with Implications to Urban Planning and Design. Springer, Berlin (2012)
Prosperi, D.C., Morgado, S.: Resilience and transformation: can we have both? In: Schrenk, M., Popovich, V.V., Zeile, P. (eds.) Proceedings of REAL CORP 2011, 819–829 (2011)
de Roo, G., Hillier, J., Van Wezemael, J. (eds.): Complexity and Planning: Systems, Assemblages and Simulations. Ashgate, Burlington (2012)
de Roo, G., Silva, E.A. (eds.): A Planner's Encounter with Complexity. Ashgate, Farnham (2010)
Schmidt, J.A.: Städtebau und evolutiver Strukturwandel. Lang, New York (1990)
Schmidt, J.A., Walloth, C.: Die Stadt als komplexes System: Urbane Anpassungsfähigkeit und Resilience. RaumPlanung **164**, 14–18 (2012)
Schultz, R., Wagner, H.-J. (eds.): Innovative Modellierung und Optimierung von Energie systemen. LIT, Berlin (2009)
Vester, F.: Urban Systems in Crisis: Understanding and Planning Human Living Spaces: The Biocybernetic Approach. Deutsche Verlags-Anstalt, Stuttgart (1976)
Vester, F., von Hesler, A.: Sensitivitätsmodell. uvf, Frankfurt am Main (1980)
Weaver, W.: Science and complexity. Am. Sci. **36**(4), 536–544 (1948)
Wehling, H.-W.: Organized and Disorganized Complexities and Socio-Economic Implications in the Northern Ruhr Area. In: Walloth, C., Gurr, J.M., Schmidt, J.A. (eds.) Understanding Complex Urban Systems: Multidisciplinary Approaches to Modeling, pp. 87–101. Springer, Cham (2014)
Weidlich, W.: Quantitative social science. Phys. Scr. **35**, 378–380 (1987)

Multimethod Modeling and Simulation Supporting Urban Planning Decisions

Ernst Gebetsroither-Geringer

Abstract This chapter reflects on why multimethod simulation is gaining increasing numbers of supporters. The chapter illustrates advantages and disadvantages of combining different modeling methods and presents a specialized software tool—the MASGISmo simulation platform. The theoretical discussion is supported with results obtained from different simulation projects. The chapter argues how an urban development model using a multimethod approach can support policy makers and urban planners in implementing robust and better acknowledged planning measures. Agent-based modeling (ABM) is used to enable model users to interpret and react to information from different levels of the urban spatial hierarchy within the simulation. The contribution also points out the value added by combining ABM with system dynamics modeling along with the use of geographical information system data. Finally, the chapter discusses how a new way of real stakeholder interactivity within the simulation can be achieved in order to improve the model.

Keywords Multimethod-modeling · Simulation platform · Agent-based modeling · Multi-agent system modeling · System dynamics modeling · Geographic information systems · Urban planning · Spatial simulation · Decision support system

Evolving regions or cities are often based on an interaction between top-down planning decisions and bottom-up processes. This interaction allows stable structures to develop, with a complex organization and a connectivity-rich network (Salat and Bourdic 2012, p. 60). Owing to the high complexity present on and between various spatial and hierarchical levels, computer models have proven useful in the analysis of different urban developmental paths. Complexity in this

E. Gebetsroither-Geringer (✉)
Energy Department, Austrian Institute of Technology, Giefinggasse 2, 1210 Vienna, Austria
e-mail: Ernst.Gebetsroither@ait.ac.at

C. Walloth et al. (eds.), *Understanding Complex Urban Systems: Multidisciplinary Approaches to Modeling*, Understanding Complex Systems,
DOI: 10.1007/978-3-319-02996-2_2,

context means a (non-linear) feedback structure connecting the elements within one and on different levels of the system.

Urban modeling was defined by Batty as follows:

> The process of identifying appropriate theory, translating this into a mathematical or formal model, developing relevant computer programs and then confronting the model with data so that it might be calibrated, validated and verified prior to its use in prediction. (Batty 1976, p. 3)

Two commonly used modeling methods, system dynamics (SD) and agent-based modeling (ABM), will be introduced in the following paragraphs. Subsequently, usage of data from geographical information systems (GIS) shall be presented, showing how it can be used to enhance the model. Finally, the potential integration of local stakeholders to improve the model is shown.

1 Urban and Regional Modeling Methods

1.1 SD Modeling

The fundamentals of SD modeling were determined by Jay Wright Forrester in the mid-1950s. SD modeling is a method that allows the understanding of the behavior of complex systems over time.

The System Dynamics Society offers the following definition:

> System dynamics is a computer-aided approach to policy analysis and design. It applies to dynamic problems arising in complex social, managerial, economic, or ecological systems—literally any dynamic systems characterized by interdependence, mutual interaction, information feedback, and circular causality. (System Dynamics 2012)

Feedback serves as the differentiating descriptor in this context. Feedback refers to the situation of X affecting Y and Y in turn affecting X, possibly through a chain of causes and effects. Furthermore, complex systems are driven by more than one simple feedback loop—they commonly include positive-feedback loops involving exponential growth processes interacting with negative, goal-seeking loops.

So, in general, it can be stated that SD modeling describes complex systems through the use of feedback loops, stocks and flows. Stocks characterize the state of the system. They are the 'memory' of a system and enable us to describe the current status of a system. Flows affect the stocks via inflow or outflow and interlink the stocks within a system. The resulting structure of the system, built up with stocks and flows, determines the behavior of the system. The following Fig. 1 depicts a simple Stock-Flow example. In principle, each SD model is built up with these building blocks. Notice feedback structures are included, e.g. between Inflow of Stock A and Stock A itself.

Together with John Collins, Forrester worked on SD modeling of urban systems and published *Urban Dynamics* in 1969 (Forrester 1969). One of the main reasons

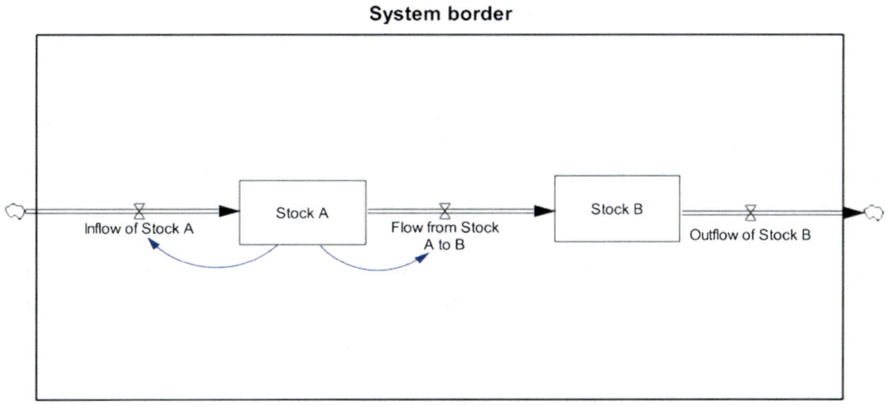

Fig. 1 Simple stock-flow example. © AIT

their collaboration focused on urban systems was Forrester's discovery that cities behave counter-intuitively. For example, corrective action suggested by urban planners is often ineffective or worsens a problem, because a simple cause and effect connection is ignored. Forrester was surprised by the success the book enjoyed and noted that "Urban Dynamics was the first of [his] modeling works that produced strong, emotional reactions" (Forrester 1995, p. 8).

The process of model building was, from the very beginning, supported by specialized software, which was one main reason why this method has become widespread over the last few decades. Today, a variety of tools exist, such as Stella/iThink (Stella 2012), Powersim (Powersim 2012) and Vensim (Vensim 2012).

Despite the benefit of understanding the behavior of complex feedback systems with simple 'building blocks' like stocks and flows that the SD method provides, there are major disadvantages. For example, emergent spatial development—i.e. the self-organized development of new and coherent structures, patterns, and properties (Goldstein 1999, p. 49)—cannot be modeled, since new stocks and flows cannot be generated during a simulation. However, emerging patterns are an important factor of sustainable regional development (cf. Salat and Bourdic 2012, p. 11). Therefore, a combined method that offers the advantage of SD modeling as well as the ability to analyze spatial development and emerging properties can help to improve regional (urban) modeling. Such a method is ABM.

1.2 Agent-Based Modeling

ABM, also sometimes called individual-based modeling (IBM) or multi-agent systems modeling (MAS), has gained increasing importance in the studies of social and economic systems. It has often been used to improve the understanding of a wide range of problems and to help forecast the effects of top-down decisions on the micro-level. Applications include the emergence of cooperation (Holland and Miller 1991) and the influence of expectations, e.g. on the stock market (Axelrod 1997a, b).

A famous early example of ABM used in urban modeling concerned the emergence of racial segregation in cities (Schelling and Hamburger 1979). However, only over the last five to ten years has ABM been receiving increased attention from the spatial development modeling community (land-use modeling as well as urban planning). It has been recognized that ABM offers a way of incorporating the influence of human decision making on land-use in a formal and spatially explicit way, taking into account social interaction, adaptation, and decision-making on different spatial and (or) hierarchical levels (Matthews et al. 2007, p. 1448).

In contrast to SD models, which are composed of stocks and flows, the building blocks of ABM and in particular the concept of agents itself are not clearly defined. However, it is argued by Jennings et al. (1998, p. 8) that ABM uses three key terms: 'situatedness', 'autonomy', and 'flexibility'. Here, 'situatedness' means that an agent receives information about the environment from sensors and, subsequently, can perform actions, which, in turn, can influence the environment. 'Autonomy' means that an agent can act solely based upon its objectives and the system's internal state, without any direct external influence. 'Flexibility' means that the agent has the ability to change its behavior, for instance when it needs to adapt or learn from others. Hence, in summary we can say that agents are situated in and interacting with their environment and are capable of changing their behavior to reach their individual objectives.

ABM has several disadvantages compared to SD modeling, e.g. a higher need of data to calibrate. Moreover, ABM produces results that are more difficult to evaluate. This causes a much higher effort for model validation and verification (model evaluation) as for instance Fagiolo et al. (2006) and Werker and Brenner (2004) discussed. Although it does not always make sense to use ABM, it is especially the ability to analyze spatial development and social behavior that makes ABM very valuable in the context of regional (urban) development models (cf. Fig. 2). ABM enables us to investigate and understand patterns emerging out of self-organization among individual agents, which leads to, e.g. segregation within regions or cities.

Figure 2 depicts two different types of agents and their behavior according to their preference for living close to their own type. The model is based on the work of Schelling and Hamburger (1979) about social systems.

2 Finding the Best Modeling Method

In general, Lorenz and Jost argue that modelers often overlook other modeling methods, simply because they cannot "differentiate and apply alternative methods" which differ from the ones they are familiar with. Often, it is noticeable that

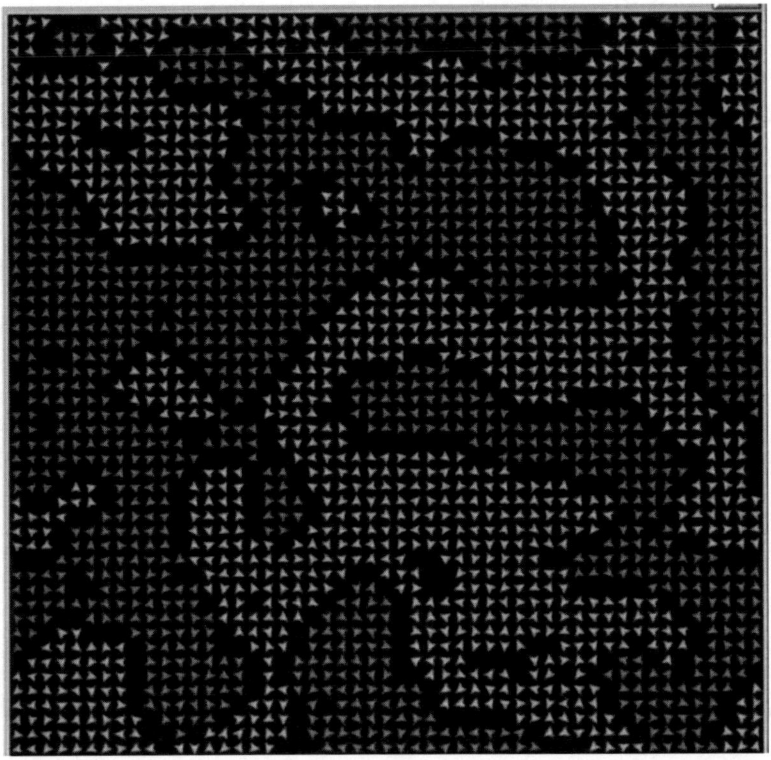

Fig. 2 Agent-based models are spatially explicit and allow us to visualize emerging patterns in cities and regions. Screenshot of the segregation model made with Netlogo © Wilensky (1997 and 1999)

modelers adopt a too strict methodological stance even where the combined use of different methods would be more appropriate (Lorenz and Jost 2006, p. 2).

Furthermore, they argue that different modeling methods would be easier to apply if software tools were readily available which offer the combined use of different modeling methods for so-called multimethod modeling. For example, the dissemination of SD modeling and ABM was significantly supported by the availability of specifically designed software tools, some of which greatly enhanced the usability of both methods (e.g., Vensim/Stella for stock flow and causal loop diagrams for SD modeling and Repast/Netlogo/Anylogic (Repast 2012; Netlogo 2012; Anylogic 2012) for ABM). In the last five to ten years, several different multimethod modeling software tools have been released and it can be assumed that this has not come to an end.

Lorenz and Jost argue that finding the most appropriate modeling method rests on the clarification of the object to be modeled, the modeling method to be applied, and the modeling purpose (Lorenz and Jost 2006). In this context, the object to be modeled stands for the system under investigation, the "what to simulate". The modeling method stands for a standardized combination of techniques and tools (Lorenz and Jost 2006, p. 3). For example, the SD modeling method stands for the technique of working with causal loop and stock flow diagrams and the use of specialized software tools (e.g. Vensim/Stella). On the other hand, the ABM method stands for the technique of defining agents through individual rules of behavior and the use of specialized software tools (e.g. Repast/Netlogo). The representation of the object to be modeled through the data available as, e.g., an object in a world of static structures or in a world of dynamic changes, as well as the detail of the data largely influence the selection of the modeling method. The purpose is the motivation behind any modeling attempt and, hence, a major factor in selecting the modeling method. For example, the purpose of modeling for policy-makers might be to solve a problem by establishing top-down measures, influencing social behavior and emergent phenomena. Such a purpose calls, on the one hand, for a modeling of the macro-level on which top-down measures can be exerted and, on the other hand, for a modeling of the micro-level on which individual actions might give rise to emergent phenomena visible on the macro-level.

Whenever two or more such levels can be distinguished in the object to be modeled, multimethod modeling might yield better results than one modeling method alone. However, it is to be considered that applying a familiar modeling method might still be better than adopting the latest one without fully understanding it. Also, it is to be considered that multimethod modeling does not only have advantages, as model evaluation often becomes more difficult with the number of approaches combined (Barlas 1989; Windrum et al. 2007; Fagiolo et al. 2006).

3 MASGISmo: A Multimethod Modeling Tool

In the beginning of ABM, spatial modeling of an agent did not include geographic information. The same was the case in the beginning of combined SD modeling and ABM (Gebetsroither 2009). Geographic information is, however, important in the simulation of, e.g., regional development, especially if local stakeholders are involved in the discussion of the result: Geographic information may enable local stakeholders to intensify their engagement in the discussion of simulation results. Therefore, especially when local stakeholders (e.g., within a participatory urban planning process using modeling) are involved, the inclusion of data from geographic information systems (GIS) represents a major advancement. Nowadays, people are used to easily accessible geographic data, thanks to ubiquitous services such as Google Maps or Open Street Maps.

Today, multimethod modeling including GIS data is possible through specifically designed software tools like Anylogic Netlogo, Repast Symphony, and MASGISmo (Multimethod Agent-based, System Dynamics and GIS modeling platform (MASGISmo 2012)). MASGISmo makes use of GIS data for complex spatial analyses, while the other software tools use their GIS functionality mainly for obtaining information about the agent's location. MASGISmo in turn enables users to analyze the environment of an agent in manifold ways within the platform, e.g., the location can be used to estimate its influence on the agent's behavior. It was developed at the Austrian Institute of Technology (AIT), the author's affiliate institution, especially to enable multimethod modeling.

MASGISmo combines SD modeling, ABM, and GIS data analyses. Combining SD and ABM is based on the pioneering works of Akkermans and Scholl (Akkermans 2001; Scholl 2001a, b) and Pourdehnad, Schieritz and Milling (Pourdehnad 2002; Schieritz and Milling 2003; Schieritz and Groessler 2003). MASGISmo combines SD modeling through inclusion of Vensim and ABM through inclusion of Repast. Enhancing the spatial capabilities of the ABM module has enabled the inclusion of GIS data analyses within the multimethod platform. This allows users of MASGISmo to develop multimethod models simulating spatially explicit actions of agents changing land-use and to spatially analyze the results of these actions. The calculation of new geographic maps out of the existing ones can be performed by using simple arithmetic operations and the agents' spatial movements by transforming land-use of single cells into steady land-use transitions. This process is part of the spatial data analysis features of MASGISmo and is one main difference to other tools like Anylogic.

The development of the simulation platform MASGISmo is predominantly determined by the requirements of the projects it serves, i.e. the objects to be modeled and the modeling purposes. Almost with every model built up with MASGISmo, new functionalities for the platform are developed, serving other future modeling purposes.

The screenshot below presents the graphical user interface (GUI) of one MASGISmo simulation showing some results (Fig. 3). Three main parts characterize MASGISmo's GUI: first the general simulation controls, second the interactive toolset and third the illustration tools such as dynamic results map, GIS layer legend and the overview map. This depicted GUI is, on the one hand, an example of the current stage of MASGISmo's development while, on the other hand, it was explicitly built for the specific purpose of the simulation of different urban development scenarios. In this use case, importing GIS data of, e.g., different urban zoning plans, new infrastructure, or shopping centers and companies enables decision makers to simulate different spatially explicit development scenarios.

Besides, since multimethod modeling should enable the user to interact with and retrieve results from the models of the different integrated methods, a new interface was developed to steer the SD models (built with Vensim, running in the background) and analyze their results within the MASGISmo GUI. Further details on building models using MASGISmo are detailed elsewhere (Gebetsroither 2009, p. 63 and MASGISmo 2012).

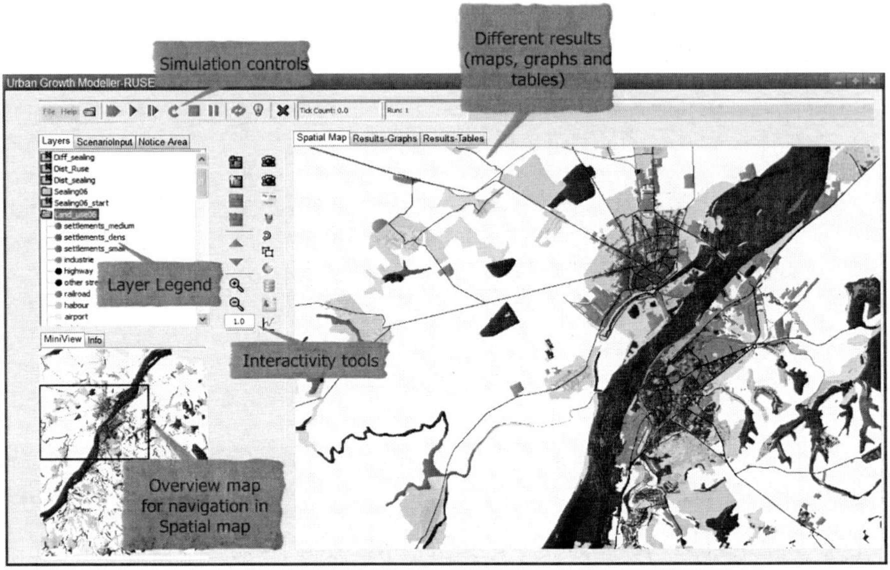

Fig. 3 Screenshot of MASGISmo's current GUI enhanced within the urbanAPI project, © AIT

4 Examples of Multimethod Modeling

In order to exemplify the use of multimethod modeling, in the following I will present the application of MASGISmo in the Dead Sea project and in the urbanAPI project. For the first project, the modeling purpose was to increase the sustainability of water management in the Dead Sea Basin. For the urbanAPI project, the modeling purpose is to support urban (regional) planning decisions with simulations and to improve regional policy making in European initiatives (European Commission 2012).

4.1 The Dead Sea Project

In the Dead Sea project two levels of objects to be modeled justified multimethod modeling. First, the region's current political framework, shaped by territorial and water claims, determined a top-down system behavior to be modeled by SD. Second, the local (spatially explicit) citizens' reactions determined a bottom-up system behavior to be modeled by ABM. The self-organization from bottom-up—expected as a reaction to the top-down political framework—would not have been possible if only SD modeling had been used.

The multimethod model was used to simulate spatially explicit future land-use scenarios, which were first introduced into the model of the region as probabilities of land-use change, based on current land-use and historical trends (Gebetsroither and Loibl 2007a). These probabilities of land-use change and the expected effects of political top-down interventions (e.g., on water prices and availability) are changing the spatial attractiveness of different areas within the entire modeled region. The changes in spatial attractiveness were then introduced into the ABM. Results of the ABM simulation included spatially explicit visualizations of future land-use scenarios, which helped to evaluate the effectiveness of political top-down interventions on local development based on defined criteria of, e.g., land availability for settlements, industrial, touristic, and agricultural activity, and natural environment (Gebetsroither and Loibl 2007b). Figure 4 depicts a scenario of land-use changes produced by the interactions between political top-down interventions and agents' activities between 2005 and 2025. The regions with significant land-use changes can be easily noticed (marked with green circles and black ellipsoids in the figure).

In the Dead Sea project, SD modeling produced the input for ABM, which in turn produced scenarios of land-use change over time, considering political top-down interventions. The SD model was used to model the physical water network with its pipelines and water storages. Also, the water demand, depending on water prices, has been simulated within the SD model. The spatially explicit results of potential land-use changes let decision makers draw conclusions on agents' reactions to interventions, test scenarios, choose those which produced more sustainable results, and to discuss regional and urban planning or economic development approaches with both public and private stakeholders. Even for the modelers and the local scientific experts, it was remarkable that, due to the use of geographic maps for the visualization of the results, new insights into the system's behavior were gained. For example, it could be depicted how a planned resettlement of about one million emigrated Palestinians would change the current land-use in the region (see the black ellipsoids in Fig. 4).

4.2 UrbanAPI Project

Another application of multimethod modeling will be developed within the urbanAPI project (urbanAPI 2012). Thereby, an urban growth model, similar to the Dead Sea project, will be established. The modeling purpose is to support policy makers' decisions on different urban development paths. Here as well, top-down policies and demographic macro-developments meet bottom-up processes and justify the use of multimethod modeling.

The urbanAPI project uses SD modeling, first, to simulate regional economic effects of top-down decisions and, second, to simulate different demographic developments. The ABM module of MASGISmo is used to simulate in a spatially explicit way the migration of households and entrepreneurs in the region around the cities Ruse (Bulgaria) and Giurgiu (Romania). This requires data on individual (agents') preferences and self-organization processes such as, e.g., social

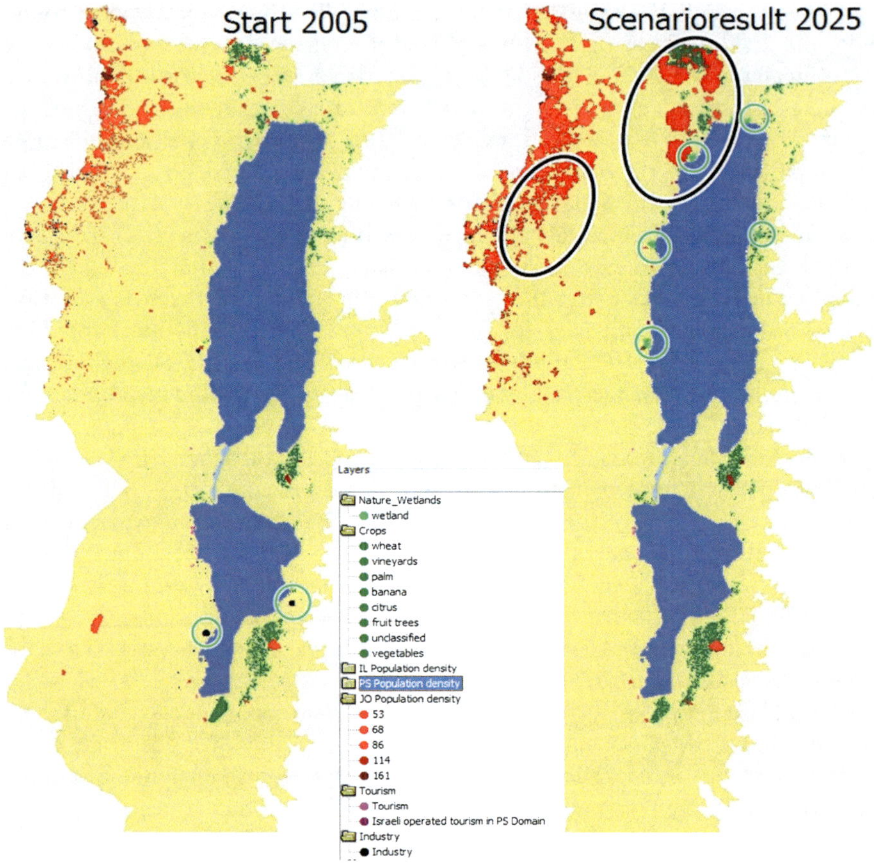

Fig. 4 Spatial development maps depicting the original land-use 2005 (*left*) and a scenario result for 2025 (*right*). Hot-spot areas are marked with *green circles* and *black ellipsoids*. A part of the legend of different land-uses is shown in-between the maps. © AIT

segregation, to be introduced into the ABM. This data—the quality of which largely determines the model's overall quality—is being drawn from the analysis of the region's historical spatial development. In the course of these historical spatial analyses, maps are produced, showing, e.g., the distances between points of interest or the household density at a certain point in time (cf. Fig. 5). From these maps, probabilities can be deduced for different agents migrating between different areas in the future and, hence, for probability maps of different types of land-use (cf. Corine Land Cover (Corine 2012)).

The four different maps in Fig. 5 show exemplarily three distance maps (in the two maps on the left the distance increases from red to green and further to blue, and in the map on the right the distance increases from green to red) and a map depicting household density (the density increasing from green to red). The model

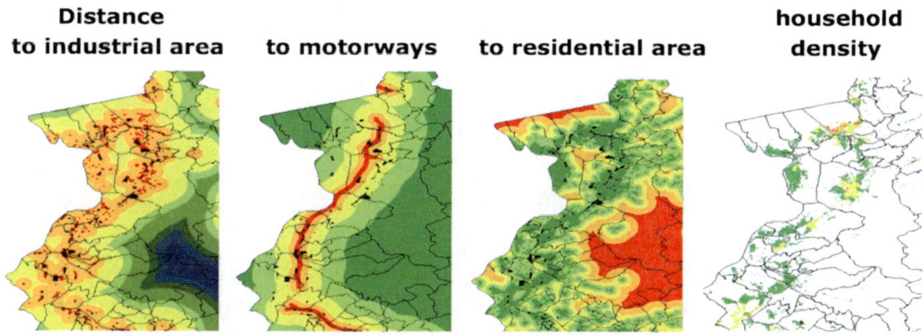

Fig. 5 Example of spatial analysis maps. © AIT

assumes that this information is used by the individual agents, according to their preferences, to decide where to move. The agents' preferences are extracted from historical data analyses revealing, e.g., where people have moved in the past in correlation to distances and densities as shown in the maps (Fig. 5). For example, data analysis showed that people moved to areas with lower household density. Further, multivariate correlation analysis showed the attractiveness of new industrial sites, influencing people's decisions where to move and, hence, impacting land-use. This kind of preference analysis is a basis for modeling different regional and urban development paths considering the individual agent's actions.

Figure 6 shows very early results of two regional development paths based on historical trends of the population development in the region and on different assumptions on the agents' preferences (in the left picture, the agents prefer to move to rural areas, whereas on the right picture the agents prefer to live in the city).

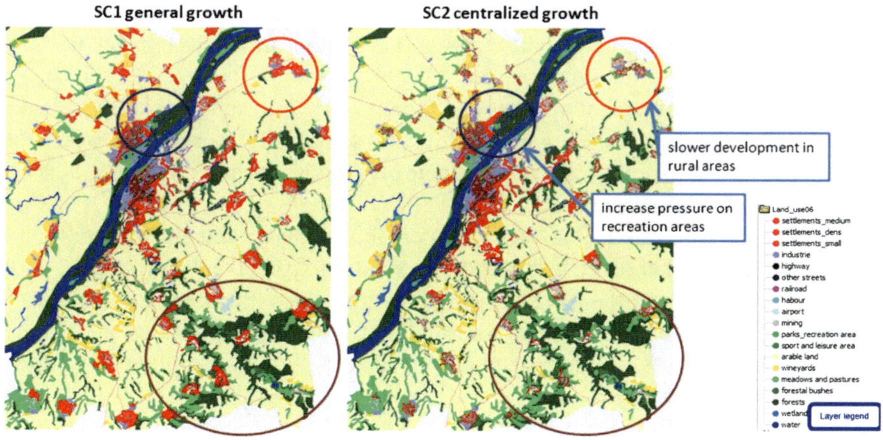

Fig. 6 urbanAPI simulation results showing two alternative urban development paths (after 20 years) for evenly distributed (*left*) and centralized (*right*) growth. The main differences between both scenarios are encircled. © AIT

5 Challenges of Multimethod Modeling in an Urban and Regional Context

The urbanAPI project, in the course of its first year, has raised a number of challenges that are symptomatic to urban and regional modeling. As mentioned above, the quality of an ABM hinges on the quality of the available data of agents' behavior.

Despite the above-mentioned thorough (multivariate correlation) analyses of the historical spatial development, an agent's behavior can only be determined to a certain extent. On the one hand, an extensive amount of historical information is necessary in order to get appropriate results. On the other hand, historical development is influenced by social conditions and political frameworks, which change over time. For example, a major change in the political framework occurred with the fall of the communist regimes and the start of the transition period in 1989/1990. Since then, people could have changed their living preferences, moving away from the industrial sites and closer to recreational areas. The analyses of the historical spatial development will show if this was really the case.

In order to reduce the impact of particular historical events in the simulation of future scenarios, a new approach is introduced. In the urbanAPI project, the author of this chapter plans to interact with local actors by using LimeSurvey, an open source survey application, over the Internet (LimeSurvey 2012). LimeSurvey will be directly connected to MASGISmo, thereby integrating the local actors' current preferences. Local actors are in this context real persons, stakeholders, whereas agents are used as artificial entities acting solely within the simulation model.

Thus, internet polls will help to refine the data on the perceived attractiveness of different areas of the region in relation to different potential urban planning scenarios which will be presented to the local actors. This means that each participant in the internet poll adds (and alters) data of the local probability layers for different land-uses and, hence, all the participants change the simulation in an iterative process. Eventually, with this method, the model should become 'better' in the sense that it reflects the local agents' actual behaviors. The behavior of the agent is dynamically changing (hence, current behavior can hardly be revealed by historical analysis), since agents can evaluate their own decisions and interact or adapt to the behavior of other agents (Benenson 2004, p. 3).

It is expected that such 'real-time' data can help urban planners and politicians to better understand the self-organization resulting from the local agents' individual decisions. The urbanAPI project will thus show the benefits of such an interactive 'real-time' approach in multimethod modeling—provided a sufficient number of local actors will take part in the surveys, which is another point to be verified by this experiment.

The interaction with local stakeholders via internet designed to improve the data on agents' behaviors is a relatively new concept that combines computer agents' and local actors' behaviors (Guyot and Honiden 2006, p. 2). It is assumed that it can increase the sustainability of urban development plans and make the citizens accept these plans more easily since it directly involves the citizens. In general, this approach aims to enable policy makers to integrate timely feedback on new infrastructure and planning guidelines into an alternative set of urban development paths. To introduce the method of (social) surveys is another step enhancing the multimethod modeling techniques beyond combining SD, ABM and GIS modeling.

6 Summary and Conclusion

Modelers such as Lorenz and Jost (2006), Akkermans (2001), Scholl (2001a, b), Pourdehnad et al. (2002), Schieritz and Milling (2003) and Schieritz and Goessler (2003) argue that we have to use combined modeling methods, because each method has its own individual research field in which it is most appropriately used. Furthermore, modeling real world phenomena needs to combine different research fields. Systems including the interaction between society and natural resources (land area can also be seen as a natural resource), determined by agents' actions that develop at different levels (micro and macro), are often better modeled using different methods.

Political decisions, for instance, take place on the macro-level, which, in turn, affects the micro-level. Often, decision makers on the macro-level (the users of the model) need to study the (self-organized) reactions of the agents on the micro-level on potential policy changes. In that way, the potential development at the micro-level influences the macro-level, the political decisions, in a feedback loop. This would not be possible if only a top-down method such as SD modeling was used.

History has shown that the development of proper simulation software was important for the development and circulation of SD and ABM (Gebetsroither 2009). Admittedly, such a development will be even more important in the case of multimethod modeling, because modelers tend to use software tools which are familiar to them, even if those tools may not be the most appropriate for their goals.

MASGISmo, a multimethod simulation platform developed over the past several years, can be used as such a tool that successfully integrates ABM, SD, as well as GIS data and analysis. The two examples provided have demonstrated how multimethod modeling can be useful in two typical cases of urban and regional development.

It is particularly in urban and regional modeling that the inclusion of individual agents' behaviors helps to better understand the system's overall behavior. The inclusion of real-time data on local agent behaviors (preferences) is now being conducted in the urbanAPI project. This makes the model come closer to the reaction of local actors and therefore it is expected to yield more realistic simulations.

Ultimately, by combining different methods, scientists with different backgrounds engage with several fundamental questions in their respective fields, such as how to build a section of an integrated model or how to parameterize and evaluate it. If these questions receive the right answers, then the strong points of each approach can be combined while their weaknesses can be mitigated.

References

Akkermans, H.: Emergent supply networks: system dynamics simulation of adaptive supply agents. In: System Sciences, Proceedings of the 34th Annual Hawaii International Conference on System Sciences, p. 11 (2001)

Anylogic http://www.anylogic.com (2012)

Axelrod, R.: Advancing the art of simulation in the social sciences. In: Conte, R., Hegselmann, R., Terna, P. (eds.) Simulating Social Phenomena. LNEMS, vol. 456, pp. 21–40. Springer, Berlin (1997a)

Axelrod, R.: The Complexity of Cooperation: Agent-Based Models of Competition and Collaboration. Princeton University Press, Princeton (1997)

Barlas, Y.: Multiple tests for validation of system dynamics type of simulation models. Eur. J. Oper. Res. **42**(1), 59–87 (1989)

Batty, M.: Urban Modelling: Algorithms, Calibrations, Predictions. Cambridge University Press, Cambridge (1976)

Benenson, I.: Agent-based modeling: from individual residential choice to urban residential dynamics. In: Goodchild, M.F., Janelle, D.G. (eds.) Spatially Integrated Social Science. Oxford University Press, Oxford (2004)

Corine, http://www.corine.dfd.dlr.de/intro_en.html (2012)

European Commission, http://ec.europa.eu/europe2020/europe-2020-in-a-nutshell/flagship-initiatives/index_en.htm (2012)

Fagiolo, G., Windrum, P., Moneta, A.: Empirical validation of agent-based models: a critical survey. ILEM Papers Series 2006/14, 1–45 (2006)

Forrester, J.W.: Urban Dynamics. MIT Press, Cambridge (1969)

Forrester, J.W.: The beginning of system dynamics. McKinsey Q. **4**, 4–16 (1995)

Gebetsroither, E.: Combining multi-agent systems modelling and system dynamics modelling in theory and practice. Alpen-Adria Universität Klagenfurt: Fakultät für Technische Wissenschaften (2009)

Gebetsroither, E., Loibl, W.: GIS-based water resource management of the Dead Sea region—integrating GIS, system dynamics and agent based modelling. In: Zeil, P., Kienberger, S. (eds.) Geoinformation for Development: Bridging the Divide through Partnerships, pp. 26–32. Wichmann, Heidelberg (2007a)

Gebetsroither, E., Loibl, W.: Middle East water challenge: supporting decision and negotiation processes through modelling land use change probability in the Dead Sea area considering alternative water supply policies. In: Zeil, P., Kienberger, S. (eds.) Geoinformation for Development: Bridging the Divide through Partnerships, pp. 147–153. Wichmann, Heidelberg (2007b)

Goldstein, J.: Emergence as a construct: history and issues. Emergence **1**(1), 49–72 (1999)

Guyot, P., Honiden, S.: Agent-based participatory simulations: merging multi-agent systems and role-playing games. J. Artif. Soc. Soc. Simul. **9**(4), 1–8 (2006)

Hamburger, H.: Individuals and aggregates. Science **205** (4401), 37–38 (1979)

Holland, J.H., Miller, J.H.: Artificial adaptive agents in economic theory. Am. Econ. Rev. **81**(2), 365–371 (1991)

Ithink/Stella, http://www.iseesystems.com/ (2012)

Jennings, N.R., Sycara, K., Wooldridge, M.J.: A Roadmap of Agent Research and Development. Autonomous Agents and Multi-Agent Systems. Kluwer, Boston (1998)
Limesurvey, http://www.limesurvey.org (2012)
Lorenz, T., Jost, A.: Towards an orientation-framework for multiparadigm modeling. In: Proceedings of the 24th International Conference of the System Dynamics Society, pp. 1–18 (2006)
MASGISmo, http://systemsresearch.ac.at/exchange/gebetsroither/Tutorial_von_MASGISmo/Welcome.html (2012)
Matthews, R.B., Gilbert, N.G., Roach, A., Polhill, J.G., Gotts, N.M.: Agent-based land-use models: a review of applications. Landsc. Ecol. **22**, 1447–1459 (2007)
Netlogo, http://ccl.northwestern.edu/netlogo/(2012)
Pourdehnad, J., Maani, K., Sedehi, H.: System dynamics and intelligent agent-based simulation: where is the synergy. In: Proceedings of the 20th International Conference of the System Dynamics Society (2002)
Powersim http://www.powersim.com (2012)
RepastJ: Recursive porous agent simulation toolkit. http://repast.sourceforge.net/repast_3/index.html (2012)
Salat, S., Bourdic, L.: Systemic resilience of complex urban systems: on trees and leaves. J. Land Use Mobil. Environ. **5**(2), 55–68 (2012)
Schieritz, N., Milling, P.: Modeling the forest or modeling the trees. In: Proceedings of the 21st International Conference of the System Dynamics Society (2003)
Schieritz, N., Groessler, A.: Emergent structures in supply chains: a study integrating agent-based and system dynamics modeling. In: Proceedings of the 36th Hawaii International Conference on System Sciences (HICSS'03) (2003)
Scholl, H.J.: Looking across the fence: comparing findings from SD modeling efforts with those of other modeling techniques. In: Proceedings of the 19th International Conference of the System Dynamics Society, Atlanta (2001a)
Scholl, H.J.: Agent-based and system dynamics modeling: a call for cross study and joint research. In: Proceedings of the 34th Annual Hawaii International Conference on System Sciences, pp. 1–8 (2001b)
System dynamics, http://www.systemdynamics.org/what-is-s/ (2012)
urbanAPI: The urbanAPI-Project: Interactive analysis, simulation and visualization tools for urban agile policy implementation (FP7-ICT-2011-7), http://www.urbanapi.eu/ (2012)
Vensim, http://www.vensim.com/ (2012)
Werker, C., Brenner, T.: Empirical Calibration of Simulation Models. Eindhoven Center for Innovation Studies, (ECIS) working paper series 04.13, Eindhoven Center (2004)
Wilensky, U.: NetLogo segregation model. http://ccl.northwestern.edu/netlogo/models/Segregation. Center for Connected Learning and Computer-Based Modeling, Northwestern Institute on Complex Systems, Northwestern University, Evanston (1997)
Wilensky, U.: NetLogo. http://ccl.northwestern.edu/netlogo/. Center for Connected Learning and Computer-Based Modeling, Northwestern Institute on Complex Systems, Northwestern University, Evanston (1999)
Windrum, P., Fagiolo, G., Moneta, A.: Empirical validation of agent-based models: alternatives and prospects. J. Artif. Soc. Soc. Simul. **10**(2), 8 (2007)

Uncertainty in Urban Systems: How to Optimize Decision Making Using Stochastic Programming

Rüdiger Schultz

"You know, the real problem in optimization is uncertainty."

Attributed to George B. Dantzig, one of the founders of optimization as a mathematical discipline.

Abstract Stochastic Optimization offers a methodology for mathematically rigorous and computationally efficient treatment of optimization problems whose data are uncertain, possibly with known probability distributions. Optimization problems in urban systems often contain uncertain data. Therefore, both stochastic optimization and urban systems may benefit from each other. In this chapter, we give an introduction to stochastic programming resorting to reduced mathematical formalism and an illustrating real-life application from distributed power generation.

Keywords Mathematical optimization · Urban systems · Decision making under uncertainty · Stochastic programming · Distributed power generation

Many, if not all, real-life optimization problems involve some kind of uncertainty. Reasons are manifold, for instance, imprecise measurement of technical or economic data, partial or complete lack of information. Or uncertainty just results from unpredictable events, such as seismic activity or weather conditions.

If a decision has to be made for an uncertain future, the decision must be optimized for a wide range of possible futures. Hence, no one future must be anticipated to simplify the decision-making problem, but rather the future must be explicitly considered uncertain, i.e. future events must remain nonanticipative. This nonanticipativity amounts to the condition that a single decision today must 'hedge' against uncertain future events. Thus, when fixing decision variables

R. Schultz (✉)
Department of Mathematics, University of Duisburg-Essen, Forsthausweg 2, 47057 Duisburg, Germany
e-mail: ruediger.schultz@uni-due.de

C. Walloth et al. (eds.), *Understanding Complex Urban Systems: Multidisciplinary Approaches to Modeling*, Understanding Complex Systems, DOI: 10.1007/978-3-319-02996-2_3, © Springer International Publishing Switzerland 2014

today, this must be done on the basis of the current information only and, in particular, this decision must not anticipate, i.e. not depend on, a particular future event. Often, improper handling of this information constraint is the reason for arriving at overly optimistic 'optimal' solutions to decision problems under uncertainty.

Decision making under uncertainty also applies to urban systems, whether it is in the operation of technical infrastructure or governance of social life. In particular concerning the latter, an adequate quantification is challenging due to the interplay between emotional desiderata and rational decision making with mathematical rigor. Rigor, however, is a foremost necessity and indispensable foundation not just in 'pure' mathematics, but also in applied mathematics where imprecision can have disastrous consequences.

The present article attempts to highlight the challenges and opportunities of optimal decision making under uncertainty and the mathematics to be applied. Using three examples from urban infrastructure and utilities, the proper handling of uncertainty will be discussed as a prevailing challenge. This challenge is mainly caused by pitfalls of 'gut-feeling' decision making that may result in loss of money or failure of infrastructure (Sect. 1). Thus, it is depicted that taking averages in the correct order is of paramount importance in optimal decision making under uncertainty (Sect. 2). Furthermore, stochastic optimization as a tool of computer-aided decision support for large decision problems, i.e. problems with many interdependent variables, will be introduced (Sect. 3).

While the main body of this paper resorts to reduced mathematical formalism, the Appendix complements this by pointing to the mathematics of modeling, structural analysis, and algorithm design in stochastic optimization.

Introduced by Dantzig (1955) and Beale (1955), stochastic optimization is widely studied for the optimization of processes and structures relevant in urban systems, e.g., vehicle routing (Laporte et al. 1992), supply chains (Santoso et al. 2005), project planning (Demeulemeester and Herroelen 2002), traffic and transportation (Walmus and Kver 2003), and, last but not least, energy systems (Handschin et al. 2006). However, despite having a long tradition in teaching (see Birge and Louveaux (2011), Kall and Mayer (2005), Shapiro et al. (2009)), and as a research topic, the practical application of the mathematics for decision making under uncertainty (cf. Wallace and Ziemba 2005) widely lacks routine algorithmic treatment and software implementation.

1 Taking the Wrong Average

Urban systems typically involve processes and structures whose optimal development, operation and planning require uncertainty to be taken into account. In many or even in all realistic situations the precise future conditions are unknown to the municipality upon making decisions that reach into the future. Using the example of operational challenges of urban utility services, we will see that proper optimization of decision making requires mathematical rigor and that very poor results may occur when decision makers are misled by what feels reasonable but in fact is not.

1.1 Optimization of Advance-purchases for Uncertain Demands

Representative for a wide variety of decision-making problems under uncertainty in, e.g., the urban context, let us consider at first the example of a municipal utility that wants to acquire a stock of de-icing salt before winter comes. Of course, the total salt demand d (say, in tons) in the winter to come is unknown at the moment the purchase is made at attractive advance-purchase costs $c > 0$ (say, in Euros) per unit of salt. When running out of salt during the season, it will cost a higher price of backorder $b > c$ per unit of salt. If salt is left over at the end of the season, disposal or storage entails cost of holding $h > 0$ per unit of salt. How much salt shall be purchased in advance such that the total cost F of advance-purchase, backorder, and storage becomes minimal?

The feeling of having made a reasonable and fair decision might come along with purchasing for an *average* de-icing salt demand $\hat{d}$ in advance. The average demand is, for instance, $\hat{d} = 50$ tons if the salt demand is assumed to vary equally likely between 0 and 100 tons. So why not buy as much salt as will be needed on average, i.e. why not advance-purchase 50 tons of salt? Intuitively, however, this approach raises doubts. A first doubt is about how, all of a sudden, a single 'magic number' $\hat{d}$ can capture reasonably well the actual annual *variation* of salt demands that might not at all be equally likely for any demand between 0 and an assumed maximum demand. A second doubt arises when noting that the asymmetry of costs c for advance purchase, which is smaller than the costs of backorder $b > c$, and the costs $h > 0$ for holding excess stock is not at all captured in the 'magic number' $\hat{d}$. It seems fairly reasonable that the asymmetry of costs causes asymmetry in the optimal advance-purchase.

Indeed choosing the average $\hat{d}$ can be a costly pitfall. Consider for example that holding were more expensive than backorder ($h > b$), then, even for a de-icing salt demand assumed to vary equally likely between 0 and 100 tons, the optimal advance purchase should be less than the average of 50 tons in order to reduce the risk of the more costly holding in case of a mild winter.

1.2 Optimization of Structures for Uncertain Loads

The same pitfall can be illustrated once more using the example of an optimized construction for an elevated railway track, as there are in, e.g., Berlin, Hamburg, and Paris. Here, the shape of the construction itself is the variable to be optimized under uncertain internal forces and external loads. How to design the structures of the elevated track to minimize construction costs while still standing safe under uncertain loads (of, e.g., strong winds, overloaded trains, a derailed train)?

With uncertain loading, we have a situation very much comparable to the de-icing-salt problem: The role of the advance purchase is taken by the shape (the decision to be taken under uncertainty), the uncertain forces correspond to the unknown salt demand, and the deformation of the construction under uncertain loads corresponds to the compensation of excess or lacking de-icing salt.

Assuming equally likely loads acting on the structure from various directions (indicated by the diagonal arrows in Fig. 1, right) one could, e.g., assume an average load (indicated by the vertical arrows in Fig. 1, left). Optimizing the structure for this average load yields the structure depicted left in Fig. 1. Obviously, one would never buy the structure on the left as acceptable, let alone optimal. Under uncertain loading, already the smallest deviation from the average vertical loading direction can break the structure, by buckling, for instance.

Thus, simply optimizing for an average may yield bad results. Taking the average value and doing a de-facto deterministic optimization results in a best fit for this one average value while neglecting many other uncertain possibilities. Hence, care must be taken when optimizing under uncertainty and, in particular, one should not mislead oneself by optimizing for an average.

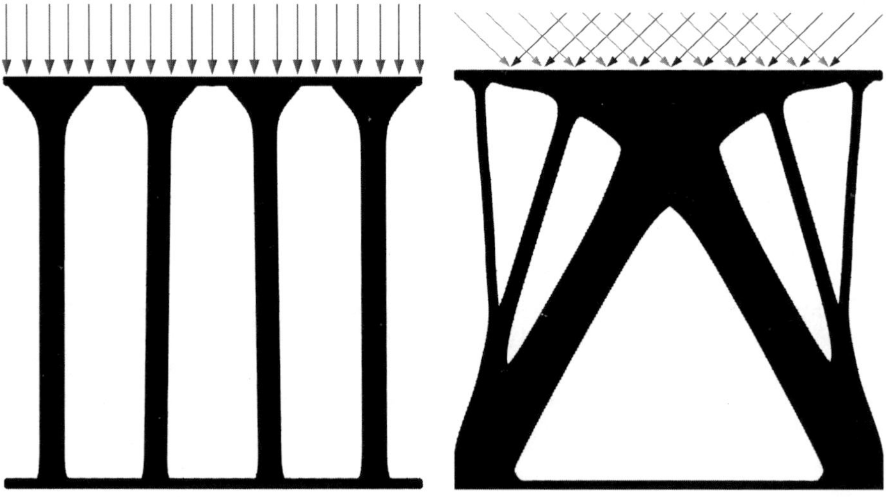

Fig. 1 The shape optimization problem depicted for the structure of, e.g., an elevated railway track, here in reduced dimension 2. The track is subjected to uncertain external load, here indicated by the diagonal arrows. On the *left*, the *vertical arrows* represent the average load, together with the resulting 'optimal' shape. On the *right*, the optimal solution is shown (see text)

2 Improved Options for Decision-making Under Uncertainty

Instead of making the final judgement depend on the average of the data, a better approach would be to make it depend on the average impact of the uncertain data on the quantities one wishes to optimze.

For example, assuming again equally likely loads acting on the structure from various directions (indicated by the diagonal arrows in Fig. 1, right), the structure can be optimized first.[1] The resulting optimal structure is depicted in Fig. 1, right.[2]

2.1 Optimize First, Then Average

For the problem of demand uncertainty of de-icing salt the analogous question is whether the total costs F can be optimized, *determining* one advance-purchase recommendation x, while considering that the costs depend on the *uncertain* demand d (instead of on the average demand $\hat{d}$). What needs to be optimized then is, hence, the cost-function $F(x, d)$ that depends on the advance-purchase x and the uncertain demand d.

In case the future demand d exceeds the advance purchase x, i.e., $d \geq x$, then

$$F(x, d) = c \cdot x + b \cdot (d - x) \tag{1}$$

with b the backorder costs. In case $d \leq x$, i.e., the future demand d is below the advance purchase x, then

$$F(x, d) = c \cdot x + h \cdot (x - d) \tag{2}$$

[1] Technically, the problem to be solved here is a shape optimization of elastic bodies with linearized elasticity under uncertain internal forces and external loads (full details of this problem are provided in Conti et al. (2008)). The deformation is governed by the Elasticity Partial Differential Equation (Elasticity PDE) that is defined on the shape and whose solution yields the displacement of the particles *inside* the body caused by the loading. Mathematical models for such shape optimizations are usually formulated in dimension three or two, the latter serving as reductions that are less complex computationally, and often already contain many important features of the three-dimensional case.—The interested reader may consult Delfour and Zolésio (2001) for details, since making the optimization problem mathematically rigorous is beyond the scope of the present paper.

[2] It should be noted, however, that such structures can be even further optimized in case the probabilities of the various possible directions of external loads are known (and not just assumed to be equally likely). In the case of a railway for example, the fixed rails determine to a large extent the directions from which loads may act upon the construction. Thus, actual structures of elevated railways may look somewhat different from the depicted example.

with h the holding (e.g., disposal or storage) costs.[3]

2.2 Optimization for Worst-case Demands

In case no assumption can be made with regard to the probability with which any future de-icing salt demand d may occur, it is totally uncertain for any advance-purchase x how close to or far off it is from the actual future demand d. In order to optimize the total costs $F(x, d)$ in this situation, the advance-purchase x that leads to the least costs in case of the worst possible over- or under-demand can be considered as optimal.

The worst-cases for any advance-purchase x in the given example occur for the maximum demand of $d_{max} = 100$ and the minimum demand of $d_{min} = 0$, respectively. These demands lead to the respective costs of

$$F(x, d_{max}) = c \cdot x + b \cdot (100 - x)$$

in case of a very snowy winter and

$$F(x, d_{min}) = c \cdot x + h \cdot (x - 0)$$

in case of a less snowy winter. Assume that backorder costs are twice as high as advance-purchase costs, i.e. $b = 2c$, and holding costs are 18 times as high as advance-purchase costs, i.e. $h = 18c$, the worst-case costs become $F(x, d_{max}) = (200 - x)c$ and $F(x; d_{min}) = 19cx$, respectively.

Mathematically, x_{opt} is at the intersection of both cost functions, i.e. where $F(x, d_{max}) = F(x; d_{min})$. $F(x, d_{max})$ *decreases* with increasing advance-purchase x (the more de-icing salt is purchased in advance, the less needs to be back-ordered in the case of maximum demand $d_{max} = 100$). $F(x, d_{min})$ *increases* with increasing advance-purchase x (the more de-icing salt is purchased in advance, the more needs to be held in case of minimum demand $d_{min} = 0$). Hence, the advance-purchase x_{opt} that is similarly costly for either worst-case—maximum demand d_{max} or minimum demand d_{min}—can be considered as optimal.

For the exemplary relation of $b = 2c$ and $h = 18c$ the optimal advance-purchase is thus $x_{opt} = 10$ tons (as compared to 50 tons for the average demand $\hat{d}$). This optimization yields worst-case costs of $F(x = 10, d_{max} = 100) = F(x = 10, d_{min} = 0) = 190c$ in case of very snowy or entirely snowless winters, and 'average case' costs of $F(x = 10, d = 50) = 90c$ (as compared to worst-case costs of $F(x = 50, d = 100) = 410c$ and $F(x = 50, d = 0) = 150c$ and 'average case' costs of $F(x = 50, d = 50) = 50c$ for an advance-purchase of an average de-icing salt demand $\hat{d} = 50$).

[3] Both terms are correct in case the advance purchase equals the future demand, i.e. $x = d$; then, $F(x, d) = c \cdot x$.

It should be noted that worst-case optimization, also called *robust optimization* in literature (cf. Ben-Tal et al. (2009)) is usually a rather conservative approximation as compared to an optimization based on an actual probability distribution $I\!P_{actual}$—that might reveal some demands as more likely, others as less likely—or even a uniform distribution $I\!P_{uniform}$—that assumes all demands to be similarly likely.

2.3 Optimization for Known Demand Distributions

For whatever amount of de-icing salt is chosen as advance purchase x, the future demand d remains the uncertain variable. For this uncertainty an underlying probability distribution $I\!P$ of possible salt demand d can be considered. This probability distribution may be based on recorded demands ($I\!P_{actual}$), or simply and hypothetically assumed as, e.g., a uniform distribution $I\!P_{uniform}$ in which each possible salt demand is considered to be equally likely.[4] For any known or assumed distribution function, the frequency information, i.e. the information about how often one particular salt demand is expected relative to any other, is given. E.g., for a uniform distribution $I\!P_{uniform}$ any salt demand d between 0 and 100 is assumed to occur equally likely.

In order to make an optimal decision better tailored to a given or assumed demand distribution, the frequency information included in any known or assumed probability distribution $I\!P$ of the de-icing salt demand d should be *explicitly* considered in the cost function $F(x,d)$. The cost function $F(x,d)$—a function of the advance purchase x and the uncertain future demand d, now assumed to occur with a likelihood given by $I\!P$—thus becomes itself an uncertain function depending on the demand's probability distribution $I\!P$. Then, in order to optimize the total costs, one may choose the minimum average total costs[5] $\hat{F}_d(x)$ that may occur for any possible advance-purchase decisions $x \geq 0$:

$$\min\{\hat{F}_d(x) : x \geq 0\}. \tag{3}$$

[4] Here 'each' relates to the famous *Problem of Measure* asking whether there exists a measure (length) assigning the value 1 to the closed interval $I = [0, 1]$, the value 0 to any singleton subset $\{\xi\}$ of I, and to **any other** subset Ξ of I a non-negative real number not exceeding one. The assignment has to be such that the measure of the union of two disjoint subsets equals the sum of the measures, and this must continue to hold for infinitly many sets, in fact, countably or enumerably many. In 1929, two pioneers of modern mathematics, Banach and Kuratowski showed that the answer is 'No', i.e., whatever assignment one takes, there always exist subsets of I not allowing for a consistent assignment in the above sense, see Dudley (1989), for instance.

[5] Note that these average costs are a function in x and no longer in d. Nevertheless, the average costs depend on the probability distribution of $d(\omega)$, justifying the notation $\hat{F}_d(x)$, with subscript d indicating dependence on its distribution.

Again, it is important to distinguish this approach of averaging the total costs $F(x,d)$ from the one of averaging the demand d introduced in Sect. 1 and shown to lead to suboptimal results. Optimizations are performed differently in both cases, i.e. while in the suboptimal approach the cost $F(x,\hat{d})$ of the average $\hat{d}$ are minimized, here the average $\hat{F}_d(x)$ of the uncertain cost is minimized. While the average $\hat{d}$ does not account for the asymmetry between backorder costs b versus holding costs h, this asymmetry is being considered when the cost function $\hat{F}_d(x)$ is optimized *before* the minimum average costs are chosen.[6]

Thus, the latter approach yields more precise results for a given or assumed demand distribution than both the approach based on an average demand $\hat{d}$ and the approach based on worst-cases.[7]

While the approach based on an average demand $\hat{d}$ for the given example suggests an advance-purchase of 50 tons of de-icing salt, the worst-case optimization suggests only 10 tons. An optimization involving a uniform distribution $\mathbb{P}_{uniform}$ suggests an even smaller advance-purchase of only 5 tons. The average total annual costs associated with these advance purchases are, in case of a uniform distribution $\mathbb{P}_{uniform}$ of de-icing salt demand over the years, $300c$ if 50 tons are purchased in advance, $100c$ if 10 tons are purchased in advance, and $97.5c$ if 5 tons are purchased in advance.[8]

3 Stochastic Optimization for Decision Making Under Uncertainty

Optimizing a function like $\hat{F}_d$ in $\min\{\hat{F}_d(x) : x \geq 0\}$ is analytically possible, i.e. can be carried out by elementary calculus, only in very specific situations (cf. Appendix A) and usually numerical, i.e. computerized, methods need to be applied. These methods, though demanding, then allow to optimize decision

[6] The latter approach comes closest to the real decision process of alternating decision and observation. It is backed by asymptotic results from stochastics, such as the Law of Large Numbers, telling under moderate assumptions that the mentioned averages converge to the 'true' (average) values with growing statistical information (for instance, growing sample sizes in estimations).

[7] It should be noted, however, that with any given or assumed distribution function the uncertainty of the decision-making problem is already being reduced before the optimization is carried out. Note also that for any of the above approaches the backorder and holding costs were assumed to be certain.

[8] This result can be obtained by minimizing $\hat{F}_d(x)$ for all possible advance-purchases $x \geq 0$, for the values $b = 2c$ and $h = 18c$ given for the de-icing salt example, and assuming a uniform distribution $\mathbb{P}_{uniform}$ of de-icing salt demands d. The interested reader may find the calculations in Appendix A.

problems with many more decision variables and uncertain parameters than in the de-icing salt problem discussed above.

'Many more', here refers to magnitudes of up to several hundreds of thousands. The methods are demanding in that their mathematical design and efficient computer implementation require specific skills at the interface of mathematics, computer science, and often the field of application the methods are applied in.

3.1 Two-stage Decision Problems Under Uncertainty

A decision problem with 'many more' uncertain parameters and decision variables arises for, e.g., the optimization of a virtual power plant (VPP, see also Handschin et al. (2006)). A VPP is a combined electricity and heat generation system consisting of many locally distributed generation units, often driven by renewable sources like wind, solar, or biomass, and being interconnected and controlled by dedicated 'smart' hard- and software. Interconnection and control makes the VPP appear to the (municipal) utility as a single larger power plant, the operation of which needs to be optimized. The optimization task can be summarized as maintaining security of supply and generating profit under uncertainty of resource availability (e.g., wind, solar irradiation, and biogas), electricity market price, and heat and electricity demand.

In order to optimize the operation of a VPP, variables forming a vector x are introduced. They reflect the unknown operational decisions and states of the system over a time horizon. The uncertain parameters are standing for uncertain load (electricity or heat demand), pool prices, and infeed from renewables. They correspond to the uncertain components of a parameter (or data) vector z.

In contrast to the simple requirement $x \geq 0$ in (3) the side conditions of the VPP problem are far more numerous. They comprise the operational, technological, and economic constraints[9] the VPP is exposed to over some time span. The latter is subdivided into quarter-hourly intervals (subperiods). Accordingly, the vector x consists of as many partial vectors as there are subperiods, i.e., 96 subperiods if wishing to optimize over a day. When doing so in a day-ahead optimization framework (as met with exchanges for power or other commodities), nonanticipativity enters as an information constraint.

The variables vector x as well as the vector z of uncertain parameters then comprise subvectors for the individual intervals. As an example, for the unknown

[9] An optimization problem asks for driving a function to its optimal (minimal or maximal) value. This may happen with or without further conditions (side conditions) on the variables the function depends on. In the former case the optimization problem is called a constrained, in the latter an unconstrained one. In a constrained optimization problem the function to be optimized is called objective function. The side conditions on the variables are called constraints, a terminus often also used for the set of vectors fulfilling the constraints.

output of a power generator, we have 96 variables (components of x) for every quarter hour of a day.

When planning in advance the optimal operation of a VPP for a period of 24 h, then, thanks to proper estimates and lasting experience, the uncertain parameters can fairly safely be assumed known with certainty for some initial part of the time span, say for the first four hours. This leads to a two-stage planning problem under uncertainty where the first-stage variables are the decisions to be taken in the initial 16 quarter-hourly intervals, and the second-stage is given by the remaining variables.

Notice the conceptual similarity with the de-icing salt problem: The role of the nonanticipative advance-purchase decisions is taken by the first-stage variables corresponding to the decisions of the first 16 quarter-hourly time intervals. The uncertain salt consumption in the de-icing model corresponds to the uncertain load, pool prices, and infeed from renewables. In the planning problem we have explicit second-stage variables, namely the decision and state variables from the seventeenth time interval on. In the de-icing salt problem the second stage is a bit hidden. In fact, it lies in the amounts of compensation after realization of the uncertain demand, i.e., either the backorder or the holding quantities.

3.2 Sizes of Real-life Decision Problems Under Uncertainty

In realistic VPP set-ups the size of a day-ahead optimization problem is large. Day-ahead planning for a VPP may be carried out, as above, in 96 quarterly-hour intervals.[10] Then, the optimization problem for an exemplary small VPP set-up of five cogeneration (combined heat and power generation) units, altogether made up of nine gas motors and one micro gas turbine, with heat storage and emergency cooler for each unit, plus eight gas boilers for peak demand of heat, one wind park with twelve turbines, and one small water power plant, involves roughly 17.500 variables and 22.000 constraints. Output variables to be decided upon are, e.g., the amount of electricity and heat to be produced, and the amount of electricity to be purchased from or sold to the market. Input parameters are, e.g., power demand, heat demand, energy prices, and wind energy availability. Among the constraints one can find, e.g., minimum and maximum power generation of each unit, maximum heat storage capacities, minimum security of supply, and minimum time to start-up the units.

The dimension, i.e. size, of the optimization problem is becoming even larger if uncertainty is taken into account. Then, the problem size can roughly be

[10] The planning horizon, in principle, can be arbitrarily long, and the planning intervals can be arbitrarily short. When wishing to arrive at meaningful results within reasonable computation time it makes a lot of sense to bound the time horizon, in our model to 24 h, because 'XXL time horizons', from certain durations on, render even the 'smartest' models intractable by even the 'smartest' algorithms. For analogous reasons the model is formulated for discrete time intervals of, e.g., 15 min, rather than in continuous time.

Table 1 Dimensions of the VPP optimization problem for different numbers of scenarios, provided for optimization problems written in so-called block-angular form (see Appendix C)

Scenarios	Boolean variables	Continuous variables	Constraints
1	8.959	8.453	22.196
5	38.719	36.613	96.084
10	75.919	71.813	188.444
50	373.519	353.413	927.324

determined by multiplying the above-mentioned number of variables and constraints by the number of scenarios for which the problem is to be solved. Here, a scenario is a single possible realization of the uncertain parameter vector z together with its (scenario) probability. In other words, there is a discrete probability distribution with finitely many realizations. Such distributions arise in two principal ways, namely as estimations from past observations or as discretizations of more complex, mainly continuous, probability distributions.[11]

Table 1 provides an impression of the problem dimension for the above-given VPP example and up to 50 scenarios.

3.3 Computational Stochastic Programming

In computational stochastic programming, the probability distribution underlying the model often is discrete, because, otherwise, one would quickly end up with numerical intractability. In stochastic programming, typically, one has to handle quite a number of uncertain parameters. Consider, for instance, the demand of a single commodity over some time horizon. The dimension of the relevant probability distribution then no longer is 1 but 96, with a day subdivided into quarter-hourly intervals, each of them with an individual uncertain parameter.

When wishing to integrate a probability density function in dimension 96, the situation is as follows. Forming Riemann sums to approximate the Riemann integral for the case that the dimension is 1 instead of 96, one subdivides the interval into a number N of subintervals. Doing the analogue in dimension 2, i.e., divide each of the two components into N subintervals, leads to N^2 periods, doing it in dimension 96 results in N^{96} elements, and thus 2^{96} if cutting each quarter-hourly demand interval into (merely) two subintervals.[12]

From the above figures it becomes clear that solving stochastic programs with continuous probability distributions of the uncertain parameters is far beyond

[11] Computers cannot compute continuous functions. Hence, the continuous distribution function has to be discretized, i.e. 'chopped' into discrete slices in order to approximate the (continuous) uncertain distributions $I\!P_z$.

[12] To confirm that this number is prohibitive recall that $2^{10} = 1024$, so 2^{96} is bigger than $2^{90} = 1024^9$ which is bigger than 10^{27}.

Table 2 Comparison of at-once solver and decomposition solver for the exemplary optimization of VPP operation

Number of scenarios	**At-once solver**	
	Computing time (sec.)	Gap (%)
5	11	0,0072
10	253	0,0099
20	1.769	0,0093
30	9.486	0,0091
40	26.286	0,0107
50	22.979	0,0129
Number of scenarios	**Decomposition solver**	
	Computing time (sec.)	Gap (%)
5	8	0,0067
10	52	0,0087
20	172	0,0088
30	364	0,0093
40	487	0,0093
50	611	0,0094

The decomposition solver consistently outperforms the at-once solver in computation time to reach the level of solution precision displayed in the 'Gap' columns. These numbers mark, measured by the relative difference of the objective function values, by how much an optimal solution to the problem still may improve the objective value of the best solution found within the time span displayed in the second column

today's numerical abilities. On the other hand, there is a well-developed methodology for solving linear stochastic programs with or without integrality requirements for individual variables, given the underlying probability distributions are discrete.

Solving those problems takes computers seconds to minutes for few scenarios, but hours for several dozen scenarios. For example, a standard at-once solver, i.e. a computer-aided algorithm used for solving the given problem, may require about 4 min. for the optimization of the VPP set-up if only 10 scenarios of the uncertain variable are considered, and about 6.5 h if 50 scenarios are considered (see Table 2, first two columns). Obviously, waiting for hours, even for a medium-term planning result, can be quite a long time.

Where the application of standard at-once solvers becomes infeasible, decomposition methods offer viable alternatives. By means of decomposition methods the original problem is split into smaller subproblems, i.e. linear programs corresponding to the individual scenarios that can be solved one after the other, instead of dealing with huge models in an all-at-once procedure. The interested reader is referred to Appendices C and D for further details on decomposition methods and related software implementation.

Typically, decomposition methods 'take over' if problem sizes are becoming large or, at the latest, when memory requirements for handling the full model become excessive. Table 2 shows a comparison between computing times of commercial all-at-once solvers and the DDSIP solver as described in Carøe and Schultz (1999) and, with accent on optimization of the VPP set-up, in Handschin et al. (2006). The DDSIP solver is especially designed to perform minimizations of uncertain linear (mixed-integer) programming problems, such as a VPP optimization, with high speed. Hence, readily available models and methods provide a basis for the development of tools for optimizing specific problems in, e.g., the urban context. The DDSIP solver has been proposed in Carøe and Schultz (1999), (cf. Appendix D). It was developed by the current author and co-workers and is accessible via the web.[13]

4 Outlook: Stochastic Programs for Urban Systems

In many situations, e.g., in the operation of municipal utilities, decisions have to be taken under uncertainty. The above examples of decision making under uncertainty suggest the use of optimization models and algorithms incorporating suitable averages, and allowing for a precise representation of the incompleteness of information under which decisions have to be made. Their range of application is wide and includes, but is certainly not limited to, decision making in the operations of urban infrastructure and utilities, e.g., the optimization of a VPP under uncertainty of power prices, power demand, and infeed from renewable resources. In fact, decision making under uncertainty is a challenge inherent in complex systems, where it occurs every day in various situations.

Possessing the knowledge of stochastic optimization enables and requires decision makers to apply them in order to save often scarce resources—human, financial, natural, etc. It should have become clear through the examples provided that decisions based on the average of the uncertain quantity—the direction and amount of load on elevated tracks, the de-icing salt consumption in winter—may yield far-from-optimal results. If, however, instead of the average the probability distribution of the uncertain data is included in the optimization model, much better results can be obtained. In decision making under uncertainty, advanced stochastic modeling can, thus, save considerable amounts of money or prevent the waste of energy, human resources, and/or the exhaust of emission (as indicated by the VPP example).

However, setting up a stochastic model that takes the probability distribution of the uncertain data into account requires some more effort as compared to setting up a deterministic model that takes only known values and/or averaged data into

[13] http://www.uni-due.de/~hn215go/ddsip.shtml or http://www.neos-server.org/neos/solvers/slp:ddsip/MPS.html.

account. In real-world applications, such as the optimization of a VPP, the number of variables and constraints is large ($>$ 10.000, or even $>$ 100.000). Solving such stochastic programming problems, therefore, requires the application of highly specialized numerical methods. A common feature of these methods is the principle of decomposition. It aims at replacing an all-at-once treatment by an iterated solution of smaller pieces the full problem is broken into. Today, decomposition-based stochastic programming solvers are available and yield good results within relatively short computing times.

The intention of the above brief treatment of computational topics is to pass the message that mathematical methodology for integrated treatment of uncertainty in an optimization context is mature enough successfully to tackle real-life decision problems. In particular, recently developed decomposition methods outperform commercial all-at-once solvers if problem dimensions are becoming large. Solvers for these problems are available, both free and commercially (see the website www.stopro.org with useful information on stochastic programming) as a scientific discipline and the people pushing it ahead.

Acknowledgments The author is indebted to Christian Walloth for his invaluable advice on bridging the gap between mathematical rigor and accessibility to a wider audience of mathematical non-experts when writing this text.

Appendices

A Closed-form Optimization for Known Distributions

There are just a few, but popular stochastic programs, such as the news-vendor problem, of which the de-icing-salt problem

$$\min\left\{\hat{F}_d(x) : x \geq 0\right\}. \tag{4}$$

given in (3) is a close cousin, allowing for closed-form solution. In the didactically reduced style of our presentation elsewhere in the paper we had referred to $\hat{F}_d(x)$ as the 'average total costs'. For the sake of mathematical precision but also to facilitate the presentation of more difficult and complex mathematical results, let us model the uncertain demand d of de-icing salt as a random variable $d(\omega)$ on a probability space $(\Omega, \mathcal{A}, I\!P)$. The average total costs $\hat{F}_d(x)$ then coincide with the following expected value[14]

$$I\!E_\omega[F(x,d(\omega))] := \int_\Omega F(x,d(\omega))\, I\!P(d\omega).$$

[14] Notice the difference between $d(\omega)$—the stochastic demand—and $d\omega$—where d symbolizes the infinitesimal differences when defining differentiation or integration, respectively.

This quantity can be computed as follows:

$$\begin{aligned}
I\!E_\omega[F(x,d(\omega))] &= cx + \int_{d(\omega)\geq x} b(d(\omega)-x)\, I\!P(d\omega) + \int_{d(\omega)<x} h(x-d(\omega))\, I\!P(d\omega) \\
&\quad + \int_{d(\omega)<x} bd(\omega)\, I\!P(d\omega) - \int_{d(\omega)<x} bd(\omega)\, I\!P(d\omega) \\
&= cx + b\, I\!E_\omega[d(\omega)] + \int_{d(\omega)\geq x} -bx\, I\!P(d\omega) \\
&\quad + \int_{d(\omega)<x} (hx - hd(\omega) - bd(\omega))\, I\!P(d\omega) \\
&\quad + \int_{d(\omega)<x} bx\, I\!P(d\omega) - \int_{d(\omega)<x} bx\, I\!P(d\omega) \\
&= b\, I\!E_\omega[d(\omega)] + (c-b)x + (h+b) \int_{d(\omega)<x} (x-d(\omega))\, I\!P(\omega) \\
&= b\, I\!E_\omega[d(\omega)] + (c-b)x + (h+b)\, I\!E_\omega[x-d(\omega)]_+
\end{aligned}$$

For the **first identity** above we employ the additivity of the integral with respect to disjoint domains of integration

$$\int_\Omega F(x,d(\omega))\, I\!P(d\omega) = \int_{d(\omega)\geq x} F(x,d(\omega))\, I\!P(d\omega) + \int_{d(\omega)<x} F(x,d(\omega))\, I\!P(d\omega),$$

use the representations (1) and (2)

$$= cx + \int_{d(\omega)\geq x} b(d(\omega)-x)\, I\!P(d\omega) + \int_{d(\omega)<x} h(x-d(\omega))\, I\!P(d\omega)$$

and, finally, add the term

$$\int_{d(\omega)<x} bd(\omega)\, I\!P(d\omega) - \int_{d(\omega)<x} bd(\omega)\, I\!P(d\omega)$$

obviously coinciding with zero.

To verify the **second identity** consider the six members of the sum on the right-hand side. The discussion of the first member is trivial, likewise for the last two members which sum up to zero. The second term on the right arises from the fourth and (partially) the second term on the left. The third term on the right is the remainder of the second term on the left. The fourth term on the right comprises the third and the last on the left.

The **third identity** verifies as follows: The first term on the right is evident, the second is the sum of the first, the third, and the last on the left. The third term on the right is the sum of the fourth and the last terms on the left.

The **fourth identity** is valid because of

$$\int_{\Omega}[x-d(\omega)]_{+}\,I\!P(d\omega) = \int_{\Omega}\max\{x-d(\omega),0\}\,I\!P(d\omega) = \int_{d(\omega)<x}(x-d(\omega))\,I\!P(d\omega).$$

On the other hand, the following holds

$$\begin{aligned}\int_0^x I\!P_{\omega}[d(\omega)\le z]\,dz &= \int_0^x\int_{-\infty}^z I\!P(d\omega)dz\\ &= \int_{-\infty}^x\int_{-\infty}^z I\!P(d\omega)dz\\ &= \int_{-\infty}^x\int_{d(\omega)}^x dz\,I\!P(d\omega)\\ &= \int_{-\infty}^x (x-d(\omega))\,I\!P(d\omega)\\ &= \int_{d(\omega)<x}(x-d(\omega))\,I\!P(d\omega) \;=\; I\!E_{\omega}[x-d(\omega)]_{+}.\end{aligned}$$

The **first identity** holds in view of

$$I\!P_{\omega}[d(\omega)\le z] = \int_{-\infty}^z I\!P(d\omega).$$

For the **second identity** we have used that the demand $d(\omega)$ is nonnegative, and hence $I\!P_{\omega}[d(\omega)\le z]=0$ for $z<0$. The **third identity** follows from Fubini's Theorem about interchanging the order of integration. Indeed, with z taken as abscissa, the two-dimensional region the integral

$$\int_{-\infty}^x\int_{-\infty}^z I\!P(d\omega)dz$$

is taken over can be seen as a cone with vertex in (x,x) and generators $(-1,-1),(0,-1)$.[15] The integral then is taken column-wise: For any Z with $z\le x$ a vertical stripe with infinitesimal width bounded by $-\infty$ and z contributes to a sum whose limit for the width tending to zero finally becomes the value of the integral. Interchanging the order of integration then is nothing but working row-wise: For any $d(\omega)$ with $d(\omega)\le x$ a horizontal stripe between $d(\omega)$ and x takes the role of the vertical stripe resulting in

$$\int_{-\infty}^x\int_{d(\omega)}^x dz I\!P(d\omega).$$

[15] Alternatively, a vector belongs to this set if and only if it arises by taking a non-negative linear combinations of the generators and adding the vector (x,x).

The **fourth identity** results from calculating the inner integral, and the **final two identities** hold for the same reasons as with the final identities in the first block of equations. Altogether, we obtain

$$\mathbb{E}_\omega[F(x,d(\omega))] \;=\; b\mathbb{E}_\omega[d(\omega)] + (c-b)x + (h+b)\int_0^x \mathbb{P}_\omega[d(\omega) \le z]\,dz.$$

The function $H(z) := \mathbb{P}_\omega[d(\omega) \le z]$ is the cumulative distribution function (cdf) of d. If H is continuous, as we assume at this place for avoiding further technicalities, then $\int_0^x H(z)\,dz$ is differentiable, and $\frac{d}{dx}\int_0^x H(z)\,dz = H(x)$. The derivative of $\mathbb{E}[F(.,d)]$ then vanishes if and only if

$$H(x) \;=\; \frac{b-c}{b+h} =: \kappa. \tag{5}$$

An optimal solution[16] to (4) thus is given by $\hat{x} = H^{-1}(\kappa)$, also called the κ-quantile of d.

B General Linear Program with Uncertainties

It is rather the exception than the rule that stochastic programs allow for closed-form solution, i.e., explicit formulae representing the optimal solutions, such as formula (5) for the de-icing-salt problem we derived in Appendix A. In the present Appendix B, we fill the gap between the specific problem (4), at the one end, and the generic two-stage stochastic programming problem at the other, see (10) below.

To this end we return to the two-stage modeling paradigm in optimization under uncertainty we employed in the paper without giving too much background information. The model concerns uncertain optimization problems of the type

$$\min\{c^\top x + q^\top y \;:\; Tx + Wy = z(\omega), x \in X, y \in Y\} \tag{6}$$

where, upon taking decisions, the scheme, in fact additional constraint

$$\text{decide on } x \;\rightarrow\; \text{observe } z(\omega) \;\rightarrow\; \text{decide on } y = y(x,\omega) \tag{7}$$

must be met.

We assume x and y are decision variables belonging to the sets X and Y which are solution sets to systems of linear inequalities, possibly involving integer requirements to components of the variable vectors x, y. By $z(\omega)$ we denote the uncertain data. Restricting presence of uncertainty to z shall keep notation simple, without oversimplifying the model itself. The subsequent discussion remains valid for problems with random q, T, W, z as well.

---

[16] As a particular consequence of $b > c$ one obtains that $x = 0$ cannot be optimal for (4), so that free (unconstrained) minimization is appropriate.

Problem (6), together with condition (7) provide ample opportunity for modeling the interplay of uncertainty and decision making in an optimization context. We had seen this with the VPP-problem and will show next that the de-icing salt problem also fits under this roof. The general model will turn out very useful algorithmically, since it supports a specific block structure of the constraints which, in turn, enables problem decomposition.

Coming back to the problem of advance purchase of de-icing salt under demand uncertainty, recall that the amount x of advance purchase induces the costs of $c \cdot x + b \cdot (d - x)$ in case of backorder and $c \cdot x + h \cdot (x - d)$ in case of holding. Putting the two together one obtains[17] total costs of

$$F(x; d) = c \cdot x \; + \; b \cdot \max\{d - x, 0\} \; + \; h \cdot \max\{x - d, 0\}. \tag{8}$$

Note that with taking the decision x and observing the demand $d(\omega)$, the knowledge becomes available whether a backorder of $d(\omega) - x$ or a hold of $x - d(\omega)$, which are mutually exclusive, is necessary. In terms of optimization models this says that an optimal solution $y_{opt} = \left(y^{+}_{opt}, y^{-}_{opt}\right)$ to

$$\min_{y^+, y^-}\{b \cdot y^+ + h \cdot y^- \; : \; y^+ - y^- = d(\omega) - x, \; y^+ \geq 0, y^- \geq 0\}.$$

must have either backorder or holding equal to zero. Indeed, when facing a deficit $d - x > 0$ it, of course, would be feasible to backorder more than $d - x$, which compensates the deficit but creates a non-zero residual amount of salt for which holding costs apply. Similarly, if $x - d > 0$, i.e., with too much salt at the end of the season, it would be feasible to even increase the overflow by a backorder, and then compensate this amount in addition to the already existing overflow. Thus, in both cases, the claim that, at optimality, either backorder or holding must be zero, has been established.

Altogether,

$$\begin{aligned} F(x, d(\omega)) &= cx + \min\{b \cdot y^+ + h \cdot y^- \; : \\ & y^+ - y^- = d(\omega) - x, \; y^+ \geq 0, y^- \geq 0\}, \end{aligned}$$

saying that $F(x, d(\omega))$ from formula (8), is a special case of

$$F(x, z(\omega)) = c^\top x + \min_{y}\{q^\top y \; : \; Wy = z(\omega) - Tx, \; y \in Y\}. \tag{9}$$

Averaging over the function values arising from the uncertain $z(\omega)$ yields a function $\hat{F}_z(x)$ which is analogous to the averaged costs $\hat{F}_d(x)$ occuring in (3). The analogue to the minimization problem or better say the generic representative arising when applying the 'optimize-then-average' principle to problem (6) under the nonanticipativity condition (7) reads

[17] The mathematical operator $\max\{\alpha, \beta\}$ turns an input (α, β) into the bigger of the two or their common value when coinciding.

$$\min\{\hat{F}_z(x) \ : \ x \in X\}. \tag{10}$$

C Block-angular Form

Provided the probability distribution underlying the uncertain data of a stochastic program is finite discrete, i.e., has finitely many mass points z_ω with probabilities $\pi_\omega, \omega = 1, \ldots, S$, and following the 'optimize-then-average' modeling principle, problem

$$\min\{\hat{F}_z(x) \ : \ x \in X\}$$

admits a special structure referred to as **block-angular** which is crucial in computations. Conceptually, the random optimal values $F(x; z)$ are computed ('optimize'), the average value is taken ('then average'), leaving a function depending on x. This function is minimized over X. Representing the averaged value as the weighted sum of the individual outcomes and probabilities yields

$$\begin{aligned}
\hat{F}_z(x) &= \sum_{\omega=1}^{S} \pi_\omega F(x; z_\omega) \\
&= \sum_{\omega=1}^{S} \pi_\omega \left(c^\top x + \min\{q^\top y : Wy = z_\omega - Tx, \ y \in Y\}\right) \\
&= c^\text{T} x + \sum_{\omega=1}^{S} \pi_\omega \min\{q^\top y : Wy = z_\omega - Tx, \ y \in Y\} \\
&= \min\left\{ c^\top x + \sum_{\omega=1}^{S} \pi_\omega q^\top y_\omega : Tx + Wy_\omega = z_\omega, \ y \in Y \right\}
\end{aligned}$$

When minimizing $\hat{F}_z(x)$ over X and writing the constraints a little more explicitly, namely in **block-angular form**, the following is obtained as an equivalent model to the initial 'optimize-then-average' model:

$$\min\left\{ c^\top x + \sum_{\omega=1}^{S} \pi_\omega q^\top y_\omega : \begin{array}{llll} Tx + Wy_1 & & & = z_1 \\ Tx & +Wy_2 & & = z_2 \\ \vdots & \ddots & & \vdots \\ Tx & & +Wy_S & = z_S \\ & x \in X, \ y_\omega \in Y, \omega & & = 1, \ldots, S \end{array} \right\} \tag{11}$$

The constraints of the above model consist of S almost identical blocks

$$Tx + Wy_\omega = z_\omega, \quad \omega = 1, \ldots, S. \tag{12}$$

Almost identical means that they differ in the data vectors on the right-hand side only. Notice that the blocks are coupled via the expressions Tx occuring in each of them. As soon as the variable x becomes fixed to some value $\bar{x}$, concatenation of the blocks is removed and decoupled equality systems of the type

$$T\bar{x} + Wy_\omega = z_\omega, \quad \omega = 1, \ldots, S$$

with the variables y_ω remain. For algorithms this decoupling is most essential, because it reduces computation to a separate handling of the blocks, instead of an all-at-once approach. However, the critical issue remaining is how to find 'good' candidates for $\bar{x}$, in other words, how to organize the iteration of x.

In stochastic programming, there are various solution methods with an algorithmic core relying on the above decomposition framework. To this end, it is crucial that among the constraints of (11) there is none containing y-variables with different indices ω_1 and ω_2 at the same time. In Appendix D another decomposition method for problems meeting this conditon and leading to a fairly comprehensive practical decision tool will be introduced.

D DDSIP

The techniques used in Appendix A to solve the de-icing salt management problem with uncertain demand, last but not least, stem from calculus in one variable, slightly advanced as far as it concerns Fubini's Theorem. It is not surprising, and underpinned by first experiences with the VPP problem in Sect. 3, that real problems from industrial practice hardly ever have only one unknown and/or are infected by just one type of uncertainty. Rather one has to be prepared to deal with hundreds, if not thousands of variables and/or dozens to hundreds of sources of uncertainty at this place.

While smaller problems (in terms of numbers of variables and/or uncertain data parameters), as the advance purchase of a single (!) commodity (de-icing salt) remain numerically solvable under inclusion of 'mild nonlinearities',[18] problems of the reported magnitude, such as optimization of VPP operation, require linearity both of the objective function and the equalities and inequalities describing the constraint sets of the optimization problems. A still feasible extension of this class in terms of model tractability (see Schultz (2002)), is to allow for integer or binary

[18] 'mild' here refers to (harsh) quantitative restriction and, even more, the presence of qualitative properties, foremostly, convexity or the benefit that optimization can be done variable by variable.

variables which enable the formulation of indivisibility of a commodity or of combinatorial constraints.[19]

The capability to deal with optimization problems of this magnitude is rooted in decomposition. In the present section we will sketch a method, serving this purpose, essentially developed in the work group of the current author, Carøe and Schultz (1999), and referred to as dual decomposition. This methodology is the driving force behind the code DDSIP[20] referred to as "Decomposition Solver" in Table 2 of the present paper.

Dual decomposition relies on rewriting (12) as follows

$$\begin{aligned} Tx_\omega + Wy_\omega &= z_\omega, \quad \omega = 1, \ldots, S \\ x_1 = \ldots &= x_S. \end{aligned}$$

Here, we pretend, for a moment, that x may be anticipative, i.e., depend on ω. However, we correct this mistake immediately by claiming the identity in the second row. Without the above formulation of nonanticipativity $x_1 = \ldots = x_S$, and with a linear objective function, the problem is perfectly separable into subproblems of the size of the initial deterministic problem. Of course, it is $x_1 = \ldots = x_S$ which hardens the problem considerably because it ties together the previously (with respect to ω) fully separable conditions in the first row.

Among the standard techniques of numerical optimization there is Lagrangian relaxation, with the following basic idea: If there are constraints making problem solution difficult then remove them and put their weighted sum with variable weights λ (called Lagrange multipliers) into the objective, making it a function of both $(x, y) := (x_1, \ldots, x_S, y_1, \ldots, y_S)$ and λ (called Lagrange function). Fix λ and minimize with respect to (x, y), taking advantage of decomposition into subproblems corresponding to the individual ω. Put a maximization with respect to λ on top of the minimization over (x, y) (resulting in the Lagrange dual problem and being the reason for calling the whole approach dual decomposition). The resulting maximal value either coincides with the optimal value of (11) or yields a fairly good lower bound. In the latter case a so-called branch-and-bound procedure can be employed to come closer and closer to the optimal value of (11). For further details see Carøe and Schultz (1999).

References

Beale, E.M.L.: On minimizing a convex function subject to linear inequalities. J. Roy. Stat. Soc. Series B (Methodological) **17**(2), 173–184 (1955)

Ben-Tal, A., El Ghaoui, L., Nemirovski, A.: Robust Optimization. Princeton University Press, Oxford (2009)

---

[19] For instance that at most 2 out of 10 items are selected, or power units get online.

[20] **D**ual **D**ecomposition of **S**tochastic **I**nteger **P**rograms.

Birge, J.R., Louveaux, F.: Introduction to Stochastic Programming, 2nd edn. Springer, New York (2011)
Carøe, C.C., Schultz, R.: Dual decomposition in stochastic integer programming. Oper. Res. Lett. **24**(1), 37–45 (1999)
Conti, S., Held, H., Pach, M., Rumpf, M., Schultz, R.: Shape optimization under uncertainty—a stochastic programming perspective. SIAM J. Optim. **19**(4), 1610–1632 (2008)
Dantzig, G.B.: Linear programming under uncertainty. Manage. Sci. **1**(3–4), 197–206 (1955)
Delfour, M.C., Zolésio, J.P.: Shapes and Geometries: Analysis, Differential Calculus, and Optimization. SIAM, Philadelphia (2001)
Demeulemeester, E.L., Herroelen, W.S.: Project Scheduling—A Research Handbook. Springer, New York (2002)
Dudley, R. M.: Real Analysis and Probability. Wadsworth & Brooks/Cole, Pacific Grove (1989)
Handschin, E., Neumann, H., Neise, F., Schultz, R.: Optimal operation of dispersed generation under uncertainty using mathematical programming. Int. J. Electr. Power Energy Syst. 28(9), 618–626 (2006)
Kall, P., Mayer, J.: Stochastic Linear Programming. Springer, New York (2005)
Laporte, G., Louveaux, F., Mercure, H.: The vehicle routing problem with stochastic travel times. Transp. Sci. **26**(3), 161–170 (1992)
Santoso, T., Ahmed, S., Goetschalckx, M., Shapiro, A.: A stochastic programming approach for supply chain network design under uncertainty. Eur. J. Oper. Res. **167**(1), 96–115 (2005)
Schultz, R.: Stochastic programming with integer variables. Math. Program. **97**(1–2), 285–309 (2002)
Shapiro, A., Dentcheva, D., Ruszczynski, A.: Lectures on Stochastic Programming. SIAM-MPS, Philadelphia (2009)
Wallace, S.W., Ziemba, W.T. (eds.): Applications of Stochastic Programming. SIAM-MPS, Philadelphia (2005)
Walmus, S., Kver, A.: Stochastic programming in transportation and logistics. In Ruszczynski, A. Shapiro, A. (eds.) Handbooks in Operations Research and Management Science, vol. 10, pp. 555–637. Elsevier, Amsterdam (2003)

Understanding Effects of Complexity in Cities During Disasters

Funda Atun

Abstract Cities are considered as complex systems which consist of highly heterogeneous and interconnected sub-systems and autonomous entities connected by non-linear, multiple interactions. When such a complex system is struck by a hazard, it stops operating as in normal conditions, due to interrupted interconnections among sub-systems. Which sub-systems and/or their interconnections are vulnerable to disasters and may create disruption to the urban system can be assessed by vulnerability analysis. However, how a disruption in a sub-system could affect the urban system cannot be foreseen in every detail prior to the occurrence of an event. In order to further the understanding of cities during disasters, this paper gives examples of consequences of uncertain interactions between urban systems and unforeseen reactions of humans during disasters such as the 1995 Kobe Earthquake, 2005 Rita and Katrina Hurricanes, and the 2010 volcanic eruption of Eyjafjallajokull.

Keywords Urban resilience · Urban complexity · Systemic vulnerability · Interdependency of systems · Natural disasters · Agent-based modeling

1 Complex Urban Systems

1.1 Cities as Complex Systems

The city is one of the largest complex spatial systems consisting of heterogeneous and interconnected elements both in physical and social structures, among them humans, organizations, infrastructures, and economy. The term 'complex' denotes

F. Atun (✉)
Department of Architecture and Urban Studies, Politecnico di Milano, Via Bonardi 3, 20133 Milan, MI, Italy
e-mail: funda.atun@polimi.it

C. Walloth et al. (eds.), *Understanding Complex Urban Systems: Multidisciplinary Approaches to Modeling*, Understanding Complex Systems,
DOI: 10.1007/978-3-319-02996-2_4,

a composite that is formed of a large number of interconnected elements. Put differently, the term 'complexity' is employed to address a great number of heterogeneous and highly interconnected elements functioning as a whole (Gell-Mann 1995). Mitchell (2009, p. 39) states that another way of understanding the meaning of 'complexity' is to characterize the features of a complex system, such as interdependency, heterogeneity and autonomy, and to observe how these features affect interactions within the system.

It must be noted, however, that the complexity of a city cannot simply be expressed in quantitative terms, e.g., by counting the large number of its elements, such as people, cars, buildings etc. Assessments of complexity must also include interdependencies between elements. Interdependency includes interactions among people and all the physical entities in the environment, such as electricity for communication systems and fuel for cars.

Arthur (1999, p. 109) defines complex systems as "process-dependent, organic and always evolving." The more they evolve, the more complex they become. A city is a good example of Arthur's definition. According to Moavenzadeh (cited in McConnell 2007, p. 25) there are five drivers of complexity of cities: "population growth, economic growth, increased urbanization, increased dependence on infrastructure and increased role of technology in society", e.g., in a positive feedback cycle, drivers like economic growth and technological advance increase the attractiveness of cities and, hence, cause further population growth. A growing population requires more land for housing—land that in many cases had been left empty for a good reason, e.g., it being prone to natural hazards. For instance, the London Dockland development increased the exposure of hazard-prone areas despite mitigation measures and, hence, increased disaster risk. Continuing growth of population, built-up structure, and infrastructure further increases the complexity of the area. The more the Docklands become interconnected with the other parts of London, the higher becomes the risk of disruption to the entire city.

In this chapter, the ultimate aim of applying concepts from complexity theory is to see the city system as a whole, to understand its complex nature with regard to disaster situations, and briefly to discuss agent-based modeling as a tool to improve disaster-resilience of urban systems.

1.2 Definition of Complexity from a Vulnerability Point of View

Occurrence of a hazard is uncertain in time, location, and/or magnitude and, consequently, it is difficult to envisage causalities, knowledge of which could help preventing damage. This difficulty increases in a complex environment such as a city, which we hardly understand and where, besides the hazards, causalities themselves appear to be uncertain. Hence, complexity is considered a main factor for hazards causing actual damage (see, e.g., White 1936; Beck 1992; Schneider

1995; Pidgeon and O'Leary 2000, p. 16) and a source of vulnerability, or, in the words of Turner et al. (2003): "Vulnerability rests in a multifaceted coupled system with connections operating at different spatio-temporal scales and commonly involving stochastic and non-linear processes".

Despite, or perhaps because of the complexity making cities vulnerable, it must be the aim of policy-makers to increase the resilience of both physical and social systems before a hazard occurs. This leads to the challenge of defining and implementing, with limited resources, an institutional context fit for an uncertain event in a highly complex environment. This, primarily, leads to the challenge of overcoming organizational deficiencies in addressing the needs of a complex system, and overcoming policies designed for static rather than dynamic systems. Indeed, cities can reduce their vulnerability by virtue of increasing complexity, e.g., by introducing feedback loops that can be used to deploy emergency-relevant information. Hence, complexity does not only amplify the effects of a disaster, it also provides opportunities to decrease the vulnerability and increase the resilience of the city.

The succeeding part of this paper explores this ambivalent role of complexity in cities. Some examples show how interdependencies of systems and human failures to comprehend the complex behavior of cities during disasters can increase a city's vulnerability. Other examples demonstrate how interdependencies and feedback mechanisms can increase the resilience of the city. The third part of this paper then returns to the challenge of overcoming organizational deficiencies and defining effective policies to increase the resilience of the complex city, discussing both the challenge itself and a potential method of addressing it through appropriate modeling of complex urban systems. Part 4 summarizes the conclusions.

2 Ambivalent Complexity of Cities

2.1 Interdependency of Systems

Kauffmann's simple model (1996, pp. 54–56), which uses buttons and strings, explains aspects like interconnectivity and interdependence within complex systems. In the model reproduced below, each two buttons are randomly either connected or not connected by a string and, as a result, all interconnected buttons generate a complex network (Fig. 1). If a button is removed from the system, then, owing to the system's interconnectivity, all other buttons will be affected.

This model raises the question of how urban systems react whenever a subsystem ceases to function properly during disasters. Indeed, in some cases, secondary and tertiary damages occur in interconnected systems. In the scientific literature dedicated to the study of disaster risk, this phenomenon is known as the 'cascade effect' or the 'domino effect'.

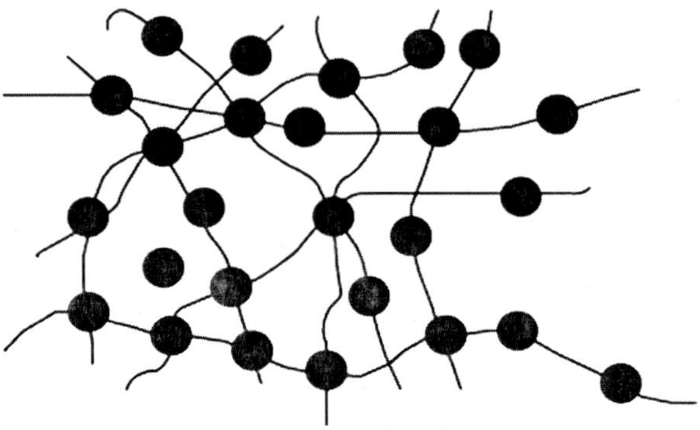

Fig. 1 Kauffmann's Button Model (after Kauffmann 1996, p. 55)

The more densely interconnected buttons in the model form clusters of high interdependency. The higher the interdependency, the more prone a system is to secondary or tertiary damages. For instance, the disruption of an electricity plant can cause further failures in the highly interdependent systems of water distribution, wastewater facilities, communication, and transportation (Fig. 2).

A case in point of how interdependency can increase a city's vulnerability is the 1995 Kobe Earthquake. It hit with a magnitude of 7.2 and resulted in 6,300 deaths, 43,177 injured, the damage or collapse of 394,440 housing units, and the destruction of 7,538 buildings (Goltz 1996).

Fig. 2 Interdependence matrix (after Paton and Johnston 2006, p. 62)

depends on ↓	Water Supply	Gas Supply	Sewage	Electricity	VHF Radio	Telephone	Roads	Rail	Air Transport	Fuel Supply	Fire Fighting
Water Supply			1								3
Gas Supply											
Sewage											
Electricity	2	1	2		2	3		1	3	2	
VHF Radio	3	3	3	3		2	2	2	3		3
Telephone	2	1	1	1	2			1	1	1	2
Roads	2	2	2	3	1	2		2	3	2	3
Rail											
Air Transport											
Fuel Supply	3	1	1		1	2	3	2	3		3
Fire Fighting									2	1	

Note: 3 = High dependence
2 = Moderate dependence
1 = Low dependence

The most severe disruptions occurred in the transportation system. The Hanshin Express Road collapsed, rail tracks were destroyed, port facilities suffered damages, while the debris caused by the earthquake filled the city's narrow streets. In particular, the collapse of the express road had a tap effect, since basic services like emergency medical care, search and rescue of the victims, and fire-fighting activities had been obstructed. The inhabitants' mobility was restricted and there were traffic jams caused by those who wanted to leave or enter the city (Goltz 1996). The city resembled an isolated island with limited accessibility from the outside, as main inland connections had been damaged.

Furthermore, organizational errors caused the failure of the automatic gas shut-down; as a consequence, the decision for a manual shut-down was taken only several hours after the earthquake (Menoni 2001). Meanwhile, owing to the gas leaks and to the traditional wooden houses, post-earthquake fires expanded and burned the overhead electric lines. The temporary loss of electricity caused a sewage release into the Osaka Bay and reduced the pumping capacity of the water distribution system, which, in turn, hindered the fire-fighting activities (Goltz 1996).

Kobe was not prepared for an earthquake, mainly because in the history of the city no major earthquake had been recorded (Shaw and Goda 2004, p. 23). The construction standards addressed other hazards known in the geographic area, like typhoons, strong winds, and landslides, while the measures of protection against earthquakes had been overlooked (Shaw and Goda 2004, p. 23). The interdependence and, sometimes, mutual exclusiveness of disaster mitigation measures cause a complex problem in itself. In Kobe, traditional wooden houses lacking interior separating walls and having heavy roofs had been built with the aim to withstand strong winds and typhoons. However, these very same features made them vulnerable to earthquakes, and, as a result, 60 % of the traditional wooden houses were seriously damaged or collapsed during the earthquake (Olshansky et al. 2006, p. 363).

It is well known that the interdependencies within complex urban systems extend spatially, beyond the city limits into larger geographical areas, and temporally, beyond the present moment into the future. Following this principle, the effects of the Kobe earthquake continued to be felt for a long period of time and spread across various areas of activity. For instance, the heavily damaged ports (combined with other damages in the transport infrastructure) led to substantial losses in the economy. This situation, in turn, led to significant changes in the social and physical structure of the city, as those who became unemployed moved to other locations within or outside Kobe. The structure of the economy also changed: after the earthquake, the heavy industry was replaced by office, service, and retail business. Furthermore, in some residential parts of the city (e.g., in the Ashiya neighborhood) the density of the residential area has increased dramatically, which made reconstruction a difficult task (Olshansky et al. 2006, p. 368).

The causalities of the 1995 Kobe earthquake clearly show how destructive an earthquake can be in a highly interdependent and densely populated system. Because of the earthquake that caused the disruption of several subsystems, the entire urban system came to a standstill. In addition to the indirect influence of the earthquake, these systems were also affected by administrative errors and systemic failures that were not visible at first sight and remained unaccounted for when later risks were assessed (see Menoni 2001).

In the case of the Kobe earthquake, overlooking the interdependencies within the complex urban system can be considered as the main cause that led to an increase in the city's vulnerability. Hence, if the interdependencies of a system are neglected, even solutions and policies aiming at solving problems may lead to new problems, as in the case of the wooden houses: on the one hand, they were protected against far more frequent disasters, but on the other hand they became vulnerable to another kind of disaster.

2.2 Human Error

Three causes of human errors can be distinguished. First, human error can be generated by one's limited cognitive capacity or knowledge, which makes decision makers misunderstand some details of a specific situation, often in a rapidly changing environment (Comfort et al. 2001, p. 144; Comfort and Haase 2006, p. 330). Second, even if well-informed decisions are made, one may not be able to implement them in time due to human and financial resource limitations. When finally implemented, the decisions might turn out to be ineffective in the complex urban system, which has meanwhile developed further. Finally, human error may be caused by insufficient dissemination of information, as well as by the citizens' neglect or misinterpretation of information (Landry and Koger 2006, p. 10).

To summarize, three causes of human errors are:

1. Limited cognitive capacity or knowledge;
2. Implementation challenges due to limited human and financial resources;
3. Insufficient dissemination of information.

There are many examples of human failures. For instance, in the Kobe case, the above-mentioned hours-long delay of a decision to close the gas valves is an illustration of the first kind of human error. The limited knowledge of the situation, the rapidly changing environment, and the failure to understand the importance of shutting down the gas valves immediately led to a great fire which increased the death toll of the earthquake. The cases of Hurricanes Katrina (August 23, 2005, category 5, Louisiana, Florida) and Rita (September 24, 2005, category 5, Florida, Louisiana, Texas) illustrate the second and the third kinds of human errors. Although the events had been forecasted long before and the population had been warned, still, the evacuation procedure largely failed, which can be traced back to human errors.

In the case of Hurricane Katrina, the incapacity to provide a sound evacuation plan which would have responded to the people's needs led to many casualties.

Besides, the delayed reaction to hazard warnings resulted in unorganized evacuations that caused severe traffic jams. Overall, the evacuation plan failed to provide solutions for all categories of inhabitants. While evacuation for car-drivers worked well, there was no evacuation plan for those who had no cars. Those who had been in charge of the evacuation plan neither understood nor addressed the citizens' needs or the reasons that prevented them from leaving the regions affected by the hurricane (Litman 2006, p. 13).

As for the disaster caused by the Rita Hurricane, it was clear that the initial assessment of the traffic volume had been wrong (Litman 2006, p. 13). This and fuel shortages caused severe traffic jams. While, on the one hand, the locals could not get out of the city, on the other hand, the emergency personnel and equipment were unable to enter the city because of the traffic jams (Litman 2006, p. 13).

In these cases, the vulnerability of the city increased due to human failure. Hence, if the potential for human failure is disregarded in the disaster risk management plans, it is difficult to intervene fast enough when human failure has occurred during a disaster. In the Kobe example, human failure caused fire, which engendered more casualties than the earthquake itself; in the Katrina example, people without cars got stuck in the city, while in the Rita example human failure led to traffic jams.

2.3 Feedback in Complex Urban Systems

Feedback within and across interconnected complex urban systems and their environment may lead to increased resilience and may provide self-restoration during and after disaster. Feedback emerges in complex systems where systems and sub-systems, e.g., social groups, are linked to each other and their members exchange information about situations, ongoing processes, or future actions (Johnson 2009, p. 67). For example, the exchange of information among individuals allows them to set up strategies and conduct their actions towards the system's stabilization in a dynamically changing environment. Hence, feedback can yield systems that are "capable of undergoing changes with relative ease" (Allen et al. 2001, cited in McConnell 2007, p. 28). Accordingly, feedback has to pertain to the system's internal interactions as well as to the relations with the system's environment.

An example of how, during a disaster, the urban system's complexity can become an advantage is provided by the quickly decaying ripples of economic effects after the 2010 eruption of Eyjafjallajokull volcano in Iceland. The economic effects could have been worse, had not the global transportation system been flexible, and had not strong communication systems enabled feedback among agents. During the volcano's eruption, flights were cancelled in Europe's northwestern regions, a situation which affected not only the passenger but also the freight transportation system on a global scale. However, the supply chains remained largely functional, since companies rapidly established contacts with each other and collaborated in an efficient way by using maritime and land routes instead of air transportation (Browne 2011). This example shows that the feedback mechanism established among social and economic agents combined with the flexibility of transportation networks helped businesses to continue operations despite the long-term effects of a natural disaster.

Moreover, feedback mechanisms may improve collective knowledge when a quick reaction to disasters is needed (Comfort 1999, p. 32). For instance, during the 1987 Whittier Narrows Earthquake in Los Angeles, streetlights and traffic lights failed due to loss of electric power, causing problems and tensions on the roads. These relaxed when a car driver undertook the responsibility of traffic control at a crossroad and, subsequently, other drivers who had noticed the first driver's action began to regulate traffic in other intersections. Eventually, with the help of several such "citizen traffic cops", the traffic began to move again. Thus, it became clear that, in order to achieve a "collective goal", people had to undertake spontaneous action without any "conscious direction" but only with the help of rapidly improving collective knowledge (Comfort 1999, p. 32; Comfort et al. 2001, p. 146).

In the cases of using feedback mechanism effectively during a disaster, the vulnerability of the city actually decreased due to large or even increasing complexity. Hence, the possibility of flexibly implementing workarounds in a disaster situation may actually complement the actions of active disaster management.

3 Resilient Complex Cities

3.1 The Challenge is Threefold

The above-stated cases exemplify three challenges of disaster risk management: first, the recognition of interdependencies among sub-systems and among their elements; second, the knowledge about indirect hazards which can be caused by complex interdependencies; and third, the enabling of flexible adaptation in case of disaster through information feedback.

Considering the first challenge, the case of the Kobe earthquake provides examples of how systems are dependent on each other. Likewise, cities are not independent from their environment. For instance, if a development plan suggests an area's densification while the disaster risk management plan suggests the opposite, both plans should be re-designed and either the development should be moved away or the disaster management plan should be improved.

Considering the second challenge, the cases of Rita and Katrina hurricanes provide examples of how indirect and multiple hazards can occur in complex systems. Since indirect and multiple hazards may lead to more casualties, they have to be considered explicitly in disaster risk management, e.g., when preparing hazard maps and evacuation plans.

Considering the third challenge, the cases of Eyjafjallajokull's eruption and the Whittier Narrows Earthquake provided examples of how to enable flexibility by improving communication networks and collective knowledge. Since flexibility provides instant solutions to emergent problems during a disaster, especially flexibility in the organizational structure has to be considered, e.g., when preparing an organizational framework in disaster management and evacuation plans.

All these three, and potential other, challenges of cities in disaster situations need to be addressed appropriately, if urban disaster resilience is to be increased. One possible approach lies in the modeling of complex urban systems through, e.g., agent-based modeling (ABM). The particular way in which ABM includes social behavior, individual perception and decision making, material structures, as well as interdependencies makes it a helpful tool in disaster risk management. Thus, the application of ABM might eventually lead to an institutional context appropriate to the challenges of complex urban systems, which increases the resilience of both physical and social systems before a hazard occurs.

3.2 Agent Based Modeling (ABM): A Tool for Dealing with Complexity

An agent-based model can simulate the interaction of individuals who take decisions under uncertain circumstances and have limited knowledge. Thus, it provides decision makers with a methodological tool for analyzing the outcomes of aggregate human behaviors. ABM has been largely used in environmental science, e.g., for negotiating groundwater demand management (Feuillette et al. 2003), optimizing the effectiveness of greenbelts in peri-urban settings (Brown et al. 2004), and improving forest ecosystem management strategies (Nute et al. 2004). Bonebeau (2002, p. 7281) describes the advantages of ABM as the ability to understand rapidly changing environments. Besides, among the advantages of this approach can be listed the capacity

- to describe and simulate complex systems (Davies et al. 2010; Chaturvedi et al. 2000; Courdier et al. 2002; Kurahashi and Terano 2005);
- to model individual decision-making entities and their interactions while dynamically linking social and environmental processes (Matthews et al. 2007, p. 1447);
- to provide enhanced speed and reliability for the system, and to tolerate uncertain data and knowledge (Park and Sugumaran 2005, p. 260).

3.3 The Advantages of ABM for Disaster Risk Management

With currently available modeling techniques it is, however, rather difficult to create a model reflecting the level of complexity required to face the challenges of disaster resilience in cities. However, being able to face the three challenges mentioned previously requires an understanding of the complex system and its interdependencies, e.g., the models could help to understand the nature of urban systems and act accordingly when preparing disaster risk management plans.

Nevertheless, the use of an ABM is an efficient method to analyze and simulate emergency and disaster situations, as it does not only count the elements, but also includes interdependencies among elements. Moreover, it considers social structures as well as physical structures, and can offer useful insights into human behavior during disasters.

ABM takes into account all three challenges by revealing interconnections among systems, potential human behaviors which could lead to failure in the system and to indirect hazards, and the discrepancy between what has been planned and the reality on the ground. Moreover, ABM reveals the hidden connections and the effects of agents' actions upon the environment (Cavalcante et al. 2012, p. 4840). Thus, it can be said that ABM provides a good picture of the possibilities arising when a situation takes a turn that is completely different from the expected one.

3.4 Development of the ABM Concept in Disaster Risk Management

In ABM, it can be noticed that the agents interact both with each other and with their environment (Fig. 3). They can also take decisions and change their actions following this interaction (Ferber 1999). In a well-delimited environment, agents collect information through their perception, or by sharing this information with other agents and acting according to the knowledge they possess. In daily routine, reactions are taken in the environment that is previously perceived by agents. However, if a disaster such as an earthquake affects a certain geographical area, the local environment will be changed. Therefore, in such a situation, any agent in the community must take action according to their perception reflecting the

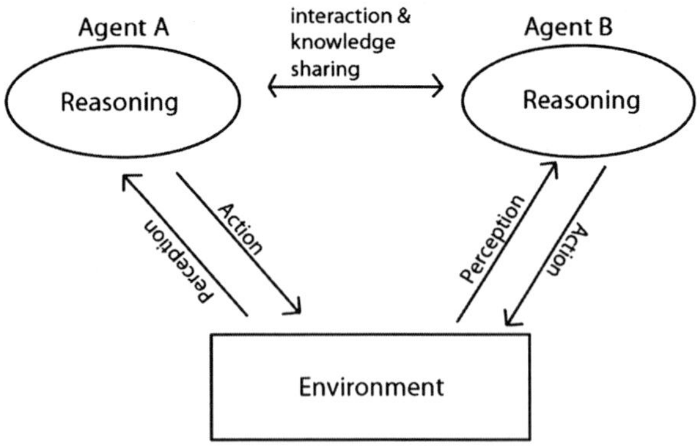

Fig. 3 A scheme of cognitive interactions between two agents and their environment (Janssen 2005, p. 4)

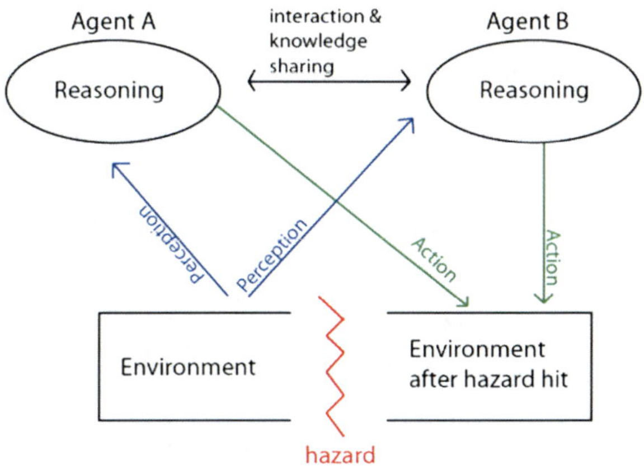

Fig. 4 A scheme of cognitive interactions between two agents and the altered environment following a disaster (scheme modified after Janssen 2005, p. 4)

elements of the pre-disaster environment. Yet, their action takes place in an environment where the circumstances are sometimes totally different (Fig. 4).

In disaster risk management, numerous agents become involved into the process of regulating operations. Generally, the connections between actions and agents easily escape one's notice. ABM helps to see those connections by simulating interaction among agents. Without knowing an agent's purpose and the quality of information he/she gathered, it is difficult to foresee his/her actions precisely before they take place. Besides, one cannot be sure that agents receive and understand all the necessary information, which, in turn, would make them decide upon the right course of action. It is clear that agents maintain multiple interaction patterns and their actions cannot be known with precision.

The interactions that can emerge among some exemplary agents are presented in Fig. 5. An agent could be any person living, working in or just passing through the area; it could be a decision maker, emergency staff or an employee of the transportation system. These agents are going to interact both with other agents and with the environment. They perceive the environment, share information, learn, adapt and act. Each agent is an autonomous entity and can choose either to follow the rules or not. An agent's behavior depends on many factors, e.g., the agent's gender, memories, religion, and education. Consequently, the result of a simulation does not always reflect the predicted aggregate behavior due to the topology of interaction.

3.5 How ABM Contributes to Urban Resilience

In their studies Hale and Heijer (2006, p. 139) state that fragmented problem-solving impedes seeing the entire system and acquiring an integrated view of all types of requirements. In addition, Hollnagel (2006, p. 12) states the difference between "normative and normal" structures of a system. "Normative structure"

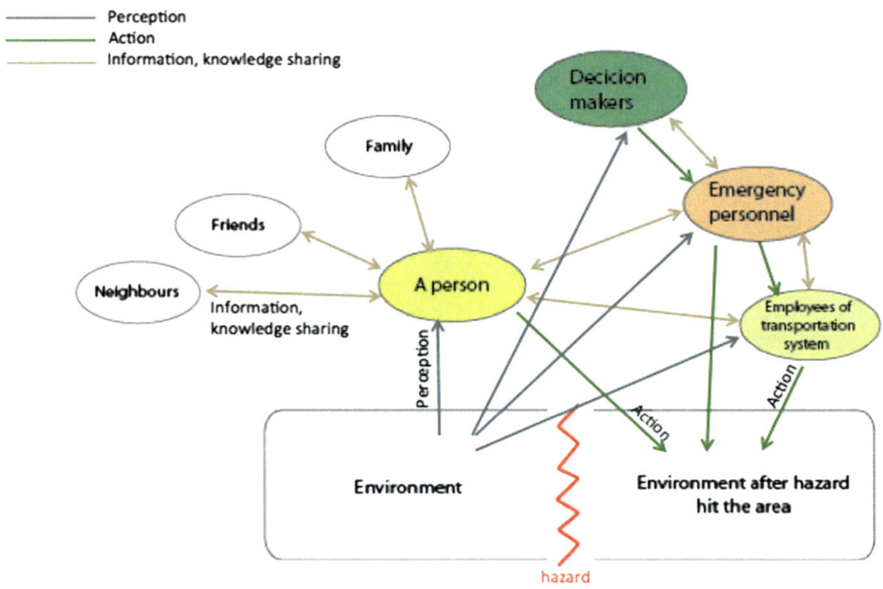

Fig. 5 Agent topology and the environment (Atun, 2013)

consists of laws, regulations and policies, and "normal structure" derives from the actual situation during a disaster, which depends on the previously given challenges. An agent-based model is a suitable representation of such a "dynamically stable" (Allen 2005, p. 18) system, by considering the original hazard as well as subsequent human failures and induced hazards. ABM helps to set the overall picture of the entire system including interconnections among elements and helps to reflect actual situations during a disaster in addition to the regulated one. Moreover, this approach contributes to resilience of the system by making interactions within the system visible, improving flexibility of the system, and consequently, by improving the ability to cope with unexpected situations.

4 Conclusion

Ultimately, the urban social structure in the city is strongly embedded within the spatial pattern of streets, buildings and other elements of infrastructure. In turn, the latter are given meaning and are shaped by their social functions. Hence, the structural systems cannot be thought of as being separated from the social subsystem. All cases show that the individual's action can make a great difference to disaster preparedness and the actual coping capacity with disaster. Or, in Kaufmann's words (1996, p. 304): "we are all part of this system, created by it, creating it." Therefore, the challenge is to provide infrastructural solutions that consider the underlying patterns of social use.

To conclude, it can be stated that, with regard to disaster risk management, the main challenge is the lack of knowledge regarding the connections among elements, within and between systems, especially between physical and social systems in urban spaces. Connected to this is the inability to predict human reactions at the moment of conceiving risk management plans. In addition to these challenges, the scale of the urban space affected by disasters and the continuously changing environment make the implementation of decisions and policies more difficult.

ABM understands agents as autonomous, social, goal-directed, and heterogeneous entities acting in a dynamic environment, whose decisions can change depending on their perception and the available information or resources. Therefore, ABM can serve as a tool for decision makers to understand potential behavioral patterns of actors during disasters, and, by the information provided in ABM simulations, disaster risk management plans can be designed to provide flexibility in structural and organizational layers to handle disaster situations.

Acknowledgement The author would like to thank Christian Walloth and Scira Menoni for their valuable comments on earlier drafts and for the support during the writing process.

References

Allen, P.M.: Cities and Regions as Self-Organizing Systems: Models of Complexity. Environmental Problems and Social Dynamics Series. Taylor and Francis, London (2005)

Allen, T., Moses, J., Hastings, D., Lloyd, S., Little, J., McGowan, D., Magee, C., Moavenzadeh, F., Nightingale, D., Roos, D., Whitney, D.: ESD Terms and Definitions (Version 12). http://esd.mit.edu/WPS/esd-wp-2002-01.pdf (2001)

Arthur, W.B.: Complexity and the economy. Science **284**, 107–109 (1999)

Atun, F.: Enhancing resilience of transportation systems in case of disasters. PhD Thesis in Spatial Planning and Urban Development Program, at Architecture and Urban Studies Department, Politecnico di Milano (2013)

Beck, U.: Risk Society: Towards a New Modernity. Sage, London (1992)

Bonebeu, E.: Agent-based modelling: methods and techniques for simulating human systems. PNAS **99**(3), 7280–7287 (2002)

Brown, D.G., Page, S.E., Riolo, R., Rand, W.: Agent-based and analytical modeling to evaluate the effectiveness of greenbelts. Environmental Modelling & Software **19**, 1097–1109 (2004)

Browne, M.: Interview with Professor Michael Browne, Professor of Logistics. University of Westminster, London (2011)

Cavalcante, R.C., Bittencourt, I.I., da Silva, A.P., Silva, M., Costa, E., Santos, R.: A survey of security in multi-agent systems. Expert Syst. Appl. **39**(5), 4835–4846 (2012)

Chaturvedi, A.R., Gupta, M., Mehta, S.R., Yue, W.T.: Agent-based simulation approach to information warfare in the SEAS environment. In: Proceedings of the 33rd Annual Hawaii International Conference System Sciences 2000 (2000)

Comfort, L.K.: Shared Risk: Complex Systems in Seismic Response. Pergamon, Oxford (1999)

Comfort, L.K., Haase, ThW: Communication, coherence, and collective action: the impact of hurricane Katrina on communications infrastructure. Public Works Manag. Policy **10**(4), 328–343 (2006)

Comfort, L.K., Sungu, Y., Johnson, D., Dunn, M.: Complex systems in crisis: anticipation and resilience in dynamic environments. J. Contingencies Crisis Manag. **9**(3), 144–158 (2001)

Courdier, R., Guerrin, F., Andriamasinoro, F.H., Paillat, J.-M.: Agent-based simulation of complex systems: application to collective management of animal wastes. J. Artif. Societies Soc. Simul. **5**(3), 1–26 (2002)
Davies, G.J., Kendall, G., Soane, E., Li, J., Charnley, F., Pollard, S.J.T.: Regulators as 'agents': power and personality in risk regulation and a role for agent-based simulation. J. Risk Res. **13**(8), 961–982 (2010)
Ferber, J.: Multi-Agent Systems: An Introduction to Distributed Artificial Intelligence. Addison-Wesley, New York (1999)
Feuillette, S., Bousquet, F., Le Goulvenc, P.: SINUSE: a multi-agent model to negotiate water demand management on a free access water table. Environ. Model. Softw. **18**, 413–427 (2003)
Gell-Mann, M.: The Quark and the Jaguar. Adventures in the Simple and Complex. Abacus, London (1995)
Goltz, J.D.: Emergency Response in the Great Hanshin-Awasi Earthquake of January 17, 1995: Planning, Mobilization and Inter-Organizational Coordination. Eleventh World Conference on Earthquake Engineering, p. 589. Elsevier, Amsterdam (1996)
Hale, A., Heijer, T.: Is resilience really necessary? the case of railways. In: Hollnagel, E., Woods, D.D., Leveson, N. (eds.) Resilience Engineering: Concepts and Precepts, pp. 125–148. Ashgate, Farnham (2006)
Hollnagel, E.: Resilience: the challenge of the unstable. In: Hollnagel, E., Woods, D.D., Leveson, N. (eds.): Resilience Engineering: Concepts and Precepts, pp. 9–18. Ashgate, Farnham (2006)
Janssen, M.A.: Agent-based Modeling. In: Proops, J., Safonov, P. (eds): Modeling in Ecological Economics, pp. 155–172. Edward Elgar, Cheltenham (2005)
Johnson, N.: Simply Complexity: A Clear Guide to Complexity Theory. One-World Publications, Oxford (2009)
Kaufmann, S.: At Home in the Universe: The Search for Laws of Self-Organization and Complexity. Oxford University Press, Oxford (1996)
Kurahashi, S., Terano, T.: Analysing norm emergence in communal sharing via agent-based simulation. Syst. Comput. Jpn **36**(6), 102–112 (2005)
Landry, B.J.L., Koger, M.S.: Dispelling 10 common disaster myths: lessons learned from hurricane Katrina and other disasters. ACM J. Educ. Res. Comput. **6**(4), 1–14 (2006)
Litman, T.: Lessons learned from Katrina and Rita: what a major disaster can teach transportation planners. J. Transp. Eng. **132**, 11–18 (2006)
McConnell, J.B.: A Life Cycle Flexibility Framework for Designing, Evaluating and Managing "Complex" Real Options: Case Studies in Urban Transportation and Aircraft Systems. Thesis of Doctor of Philosophy in Technology, Management and Policy at the Massachusetts Institute of Technology (2007)
Matthews, R.B., Gilbert, N., Roach, A., Polhill, G., Gotts, N.: Agent-based land-use models: a review of applications. Landscape Ecol. **22**, 1447–1459 (2007)
Menoni, S.: Chains of damages and failures in a metropolitan environment: some observations on the Kobe earthquake in 1995. J. Hazard. Mater. **86**, 101–119 (2001)
Mitchell, M.: Complexity: A Guided Tour. Oxford University Press, Oxford (2009)
Nute, D., Potter, W.D., Maier, F., Wang, J., Twery, M., Rauscher, H.M., Thomasma, S., Dass, M., Uchiyama, H., Glende, A.: NED-2: an agent-based decision support system for forest ecosystem management. Environ. Modell. Software **19**, 831–843 (2004)
Olshansky, R.B., Johnson, L.A., Topping, K.C.: Rebuilding communities following disaster: lessons from Kobe and Los Angeles. Built Environ. **32**(4), 357–374 (2006)
Park, S., Sugumaran, V.: Designing multi-agent systems: a framework and application. Expert Syst. Appl. **28**, 259–271 (2005)
Paton, D., Johnston, D. (eds.): Disaster Resilience: An Integrated Approach. Charles C. Thomas, Springfield, Ill. (2006)
Pidgeon, N., O'Leary, M.: Man-made disasters: why technology and organizations (sometimes) fail. Saf. Sci. **34**, 15–30 (2000)
Schneider, S.K.: Flirting with Disaster: Public Management in Crisis Situations. M.E. Sharpe, Armonk (1995)

Shaw, R., Goda, K.: From disaster to sustainable civil society: the Kobe experience. Disasters **28**(1), 16–40 (2004)

Turner, B.L., Kasperson, R.E., Matson, P.A., McCarthy, J.J., Corell, R.W., Christensen, L., Eckleyg, N., Kasperson, J.X., Luerse, A., Martellog, M.L., Polskya, C., Pulsiphera, A., Schiller, A.: A framework for vulnerability analysis in sustainability science. Proc. Nat. Acad. Sci. **100**(14), 8074–8079 (2003)

White, G.F.: Notes on flood protection and land use planning. Planners J. **3**(3), 57–61 (1936)

Towards Evolutionary Economic Analysis of Sustainable Urban Real Estate: Concept of a Research Strategy Exemplified on House Price Modeling Using the Self-Organizing Map, Interviews and Field Inspection

Tom Kauko

Abstract Evolutionary economics has begun to replace—or at least complement—neoclassical economics as the most widely accepted framework for economic modeling. Evolutionary frameworks are nonlinear and iterative: they take their starting point in the divergence of routines and outcomes of business activity over time, and work on the assumption that this diversification subsequently leads to a selection-of-the-fittest mechanism. In real estate analysis the issue is about market outcome and the behavior of actors such as developers and investors. Thus, a feasible method needs to apply both real estate transaction price data and expert interviews about actor motivations and tendencies. On top of the price development target, urban sustainability is incorporated as a more qualitative target, as the aim is to compare the price development with various sustainability aspects across different locations and typical market segments. This demonstration is based on a modeling approach known as the self-organizing map (SOM).

Keywords Evolutionary economics · Real estate · The self-organizing map (SOM) · Urban sustainability

T. Kauko (✉)
Department of Geography, Norwegian University of Science and Technology—NTNU, NO-7491 Trondheim, Norway
e-mail: tom.kauko@svt.ntnu.no

C. Walloth et al. (eds.), *Understanding Complex Urban Systems: Multidisciplinary Approaches to Modeling*, Understanding Complex Systems, DOI: 10.1007/978-3-319-02996-2_5, © Springer International Publishing Switzerland 2014

1 Introduction

Real estate theory comprises two different strands: one considers real estate as an asset (real estate economics); the other defines real estate through its *physical dimensions* (traditional real estate analysis). While relatively underdeveloped areas of academic enquiry, they both have recently gained momentum: the former due to the way real estate was at the centre stage of the sub-prime crisis and the consequences thereof (from the year 2008 onwards); the latter through the widespread sustainability discourse where buildings and land use are core concepts (from c. 2000 onwards).

The developments on real estate markets and ongoing sustainability discourses raise academic concerns about the way real estate is analyzed. First, real estate economics—just as any other neoclassical economics (NCE) model of static equilibrium—is increasingly seen as insufficient for grasping the dynamic and evolutionary processes which the real estate economy (including the construction economy) is likely to be subject to as part of the global economy. Hence, a first question of interest is to which extent an evolutionary and dynamic view of the economy is applicable to real estate transactions. Second, traditional real estate analysis seems insufficient since, in the physical sense, land and buildings are increasingly subject to sustainability evaluations along environmental, social, cultural, and economic dimensions. Hence, a second question of interest concerns the sustainability of various real estate products, for instance modern residential and office blocks.

Targeting both questions of interest, the research presented here rests on the argument that economic sustainability is a matter of both—evolutionary dynamics as known in evolutionary economics and evolutionary economic geography, and sustainability criteria relevant to most, if not all, disciplines dealing with the built environment. Evolutionary economics is known from economic modeling, where it has begun to replace—or at least complement—NCE as the most widely accepted framework since the 1980s (Nelson and Winter 1982, 2002). The crucial difference between NCE and evolutionary approaches (including the Austrian school of economics) is that modeling according to the latter allows for feedback between outcome and process, and emphasizes behavioral factors and continuous diversification and selection of the fittest business activities (routines) on top of the market outcome data employed by standard NCE approaches. Sustainability criteria allow for a sustainable development evaluation of the built environment (e.g., Ratcliffe et al. 2010), thereby going beyond NCE's narrow sustainability concept of a cost and benefit balance. It can be argued that, also when evaluating the built environment, sustainable development comprises three basic dimensions: the environmental-ecologic (green buildings/developments), the social-cultural, and the economic-financial dimensions. The most common and theorized dimension is the green aspect. The other two categories are considerably less developed in terms of analytical frameworks.

An innovative aspect lies in combining the two concepts (for related research see d'Amato and Kauko 2012; Kauko 2010, 2011). An *implicit* evolutionary perspective to real estate sustainability has recently been championed by RICS, the leading global advisory body on built environment issues (see, e.g., Macintosh 2010; Ratcliffe et al. 2010). The research strategy presented here discloses evolutionary dynamics (i.e., real estate price development) and sustainability criteria *explicitly* in order to analyze in how far a price development observed can plausibly be related to a more or less sustainable development in a given place. In an exemplary implementation of the suggested research strategy, the development of real estate prices is disclosed using the *self-organizing map* (SOM) approach with statistic data on paid transaction prices, and the level of urban sustainability is identified based on expert interviews (academic, non-governmental, and public and private sector experts) and field inspection. That any rigorous link between real estate sustainability and evolutionary real estate economics can be developed is not a foregone conclusion. Hence, the main idea of this contribution is to unveil a research strategy where environmental, social, and economic sustainability concepts are scrutinized in relation to the argumentation in evolutionary economics and evolutionary economic geography. First methodological results are drawn from recently conducted research and a number of hypotheses to be tested in further research are suggested. This research strategy is expected to open new scholarly horizons of combining the two different disciplinary traditions of evolutionary economics and sustainability criteria.

2 An Evolutionary Framework for the Analysis of Real Estate Price Development and Urban Sustainability

2.1 Urban Real Estate Analysis

It can be argued that the theory of human geography (including economic and urban geography sub-disciplines) has ignored (urban) real estate since roughly the early 1990s (if not a point further back in time). This is a point that is rarely, if ever, documented. The apparent reason is that the Marxist work of the 1970 and 1980s, which had almost completely taken over human geography theorizing at the time, was later found inapt to account for global changes, thereby realizing its inbuilt contradictions and limitations. Here, ostensibly was a silent resignation; to continue this tradition would have been an embarrassment. The attention of the theorists was directed elsewhere, which, given the lack of alternative lines to pursue, led to a halt in the development of theory of real estate within geography. An interesting side-issue within this argumentation is that regulating real estate investment, rather than understanding real estate investment itself, continued to be a popular topic in Marxist inspired human geography and land use planning, and later within the rising sustainable development paradigm. At the moment, the housing crisis has woken up the theory developers from this apparent lethargy in so far as evolutionary economic geography has recognized the void and carefully taken on the challenge (see Martin 2011). One suspects that there is more to come from this position.

Intuitively, when examining changes in the urban fabric, real estate has a central role. Cities cannot be understood without a relation to real estate development and markets. Depending on the time of development and the area's current image, different parts of the city are likely to experience upward and downward developments in the value of their real estate stock. Any investment or lack thereof will either enhance the potential of that location, thereby attracting further investment and increasing the value even further, or lead to dilapidation, loss in potential, absence of investment, and further decreases in the value. It should also be noted that either trend can take a reverse course; inappropriate structures may generate a downward trend in the price movements and development activity, or/ and gentrification of a neighborhood will lead to an upward trend.

Evolutionary approaches recognize such complex dynamics that define the developments of the trajectories of events, e.g., in urban real estate markets. Some of the theory of housing market modeling has implicitly included these evolutionary mechanisms and assumptions (see, e.g., Maclennan and Tu 1996; Kauko 2006a). The main difference to an NCE approach is the feedback between outcome and process. It is assumed (and often also verified empirically) that a certain market pricing in a real estate location may be locked into an iterative process, where the various actors react to the set prices by modifying them further—increasing, decreasing or stabilizing the price levels. Such price development, moreover, is not the result of simple supply–demand equilibrium mechanisms, but rather of dynamic and complex behavioral patterns as well as institutional market constraints. In such conditions the price is unlikely to reach market equilibrium.

Evolutionary frameworks are (by definition) nonlinear and iterative: they take their starting point in the constant diversification of business activities (routines) and their outcomes over time, and the assumption of a subsequent selection of the fittest. This implies a constant heterogeneity in product ranges (which could be fostered by flexible administrative structures that enable market sensitive behavior) influenced by the changing preferences of consumers. Given the long innovation cycles of real estate products, how are heterogeneity, flexibility, and other characteristics of evolutionary economics reflected in the real estate industry? In fact, dynamic market outcome and the behavior of actors such as developers and investors are crucial in real estate economics. Thus, evolutionary real estate analysis needs to consider both transaction prices *and* actor motivations and tendencies.

2.2 Evolutionary Economics and Evolutionary Economic Geography

Commonly, evolutionary economics is traced back to the work of Nelson and Winter (1982, 2002), and builds on the traditions of the Austrian School (Josef Schumpeter, in particular) as well as Behavioral Science (notably, Herbert Simon). The intuitive appeal of evolutionary economics is that it not only considers isolated markets (as NCE does), but the society at large in which we try, with varying success, to rationalize our choices. Economic decision-making is like other kinds of decision-making: It operates under a number of constraints. Furthermore, business actions are often routines, but when the business environment changes, these routines might have to be changed as well. However, this adjustment does not happen immediately.

Dopfer (1994) on the other hand, traces the origins of evolutionary economics along a parallel line, namely to the works of the somewhat esoteric Kenneth Boulding (1910–1993). During the late 1970s and early 1980s Boulding worked on a theory with the goal of explaining economic phenomena on the basis of evolutionary principles. Boulding's main idea is that of mutation and selection in what would be, if accepted more widely, the social-economic counterpart of evolution. The key here is the notion of global irreversibility of economic processes, in the sense that the economy always is embedded in a broad context (social, cultural, political, environmental, etc.). This view is opposed to seeing the economy as a stand-alone, closed mechanism, rigid and mechanistic as postulated by NCE, to which Boulding referred as 'cookbook theory'. Dopfer concludes that, as today evolutionary economics is attracting increasing interest, Boulding's legacy is twofold: the refutation of the mechanistic NCE, and a demonstration of the relevance of evolutionary principles in explaining economic phenomena.

Evolutionary economic geography (EEG) is seen as the spatial extension of evolutionary economics. The crucial addendum to the 'traditional' economics line is the emphasis on the path-dependent nature of clustering in time and space. In this vein, Boschma and Frenken (2011) champion the operational possibilities arising from using quantitative modeling. The key point here is that early decisions and outcomes matter in a model that emphasizes change and continuity. Martin (2012) in turn revises and extends the path-dependence concept so as to use it as an umbrella concept that covers several particular evolutionary approaches. In doing so he argues that this concept is consistent with all theoretical models that explicitly incorporate history. In a generic sense, path dependence can be applied for many phenomena both stable and unstable, Martin (2012, p. 186) argues.

Martin and Sunley (2006) add to the discussion on path dependence by stressing the theoretical aspects. They also argue that in path dependence, one cannot explain decisions made deterministically—only stochastically. Essletzbichler and Rigby (2007) argue that, despite increasing support for evolutionary economics since the early 1980s, a coherent research paradigm is still missing. They also acknowledge the importance of particular additions to the main theory, such as in particular the crucial role of market niches in a continually evolving economic environment. In other words, the most successful adapters tend to represent market segments that are far differentiated from the average market or 'bulk' segments. The relevance of this notion is also purported in the real estate sustainability discourse, as Dixon (2007) notes that it is niche developers who tend to afford sustainability elements in residential as well as office markets regardless of the market cycle.

Another recurring theme in evolutionary economics is the connection between local, i.e. micro-level, and global, i.e. macro-level, analysis. While NCE focuses on either the micro or the macro level, evolutionary economics ties both levels together (Nelson and Winter 1982; Essletzbichler and Rigby 2007).

In this vein it is perhaps necessary to note that, in relation to institutional economics, another 'middle-way theory' in real estate analysis, the evolutionary approach is far more operational and better-suited to quantitative modeling (although Plummer (2007) argues that using map patterns resulting from economic growth regressions across geographical units shows the empirical plausibility of *institutionalist* theorizing). While Maclennan and Tu (1996) see the applicability of evolutionary economics to real estate analysis in housing market research, recent surveys by RICS (Macintosh 2010; Ratcliffe et al. 2010) see the applicability more widely in research on the built environment, incorporating sustainability aspects as well as information and communication technology (ICT).

While one is generally well advised to exercise caution when attempting to use nature as an analogy for individual choice-based processes within economic systems, "reasonable evolutionary explanations" might be worth pursuing (see Robson 2002). In general terms, such explanations imply that systems inevitably will change their properties irreversibly, that the generation of diversity and subsequent selection of the fittest is the driver of this change, and that knowledge and innovation play a role in this selection process. Nelson and Winter (2002), in a similar vein, argue that evolutionary economics has much to offer in terms of interdisciplinary results. While NCE still is not interested in covering such ground, evolutionary economics builds bridges to other disciplines (biology, psychology, sociology, management studies, computer science, geography, at least) and applies ideas from other realms in a cross-disciplinary fashion. This is precisely the point of the current undertaking: Building a bridge between real estate economics and concepts of the sustainable city.

2.3 Sustainable Real Estate

How then to define what a sustainable city is in real estate analysis? We could, for example, look at fifteenth-century Florence as a model. In this city, banking innovations fostered investments into the built environment which allowed housing the masses of in-movers and facilitating the start-up of various businesses; consequently, as the times were peaceful in that period, all sorts of social and cultural life were flourishing, too. Success of late medieval Florence was in fact the result of investments—not planning. (This is not to deny the logic of public policy possibilities as a secondary influence towards sustainability; for example, with tax incentives one could attempt to steer the land and property development patterns towards a dense urban environment that would save energy and enable better management of land use and infrastructure.)

Of the various sustainability dimensions, the economic one is often difficult to comprehend. Is it not the same as profitability? Definitely not, is the answer. The difference pertains to the way incentives are set and to the time-span of the analysis which is ideally looking at least two generations ahead. Economic sustainability encourages working for profits, saving them for a rainy day and reinvesting the savings wisely. The role of government regulation then is to incentivize this behavior. Consider for example the investments in the first streetlights in the seventeenth century Spanish kingdom when, at the same time, in the Polish counterpart, then another major European power, the profits were spent lavishly by the elite 5 % of the population. In general, here sustainability is easier to exemplify via obviously unsustainable cases, where incentives for ordinary citizens to work, save or reinvest are weakened, such as during the collapse of state socialism and ostensibly also amidst the current EU banking crisis (insofar as we believe that the taxpayers' money is being pocketed by 'banksters' who aim for short term gains only realizable by a limited group of insiders).

2.4 Sustainability Analysis of Real Estate Calls for an Evolutionary Perspective

The most fruitful mould for analyzing real estate activity is arguably the residential sector. As compared to the commercial sector, the residential sector is more complex with its inherent socio-cultural and behavioral elements that go beyond the explanatory potential of 'neat economic utility theorizing'. Besides, housing development, and thereby also the housing market, comprises important elements for defining a sustainable place. In what follows, using the example of some selected sustainability criteria, the connection between sustainability and evolutionary economics, how they relate to and entangle with each other, will be developed.

Housing quality, affordability, and differentiation are three sustainability criteria that can be considered pertinent in residential real estate:

A grossly substandard level of housing is unacceptable, not least for health and safety reasons. The quality—albeit largely a subjective indicator—therefore ought to develop in the same direction and with the same pace as the price level. This requirement of quality pertains to the site and building-specific attributes as well as to the characteristics of the surrounding environment, neighborhood, and the city as a whole.

Table 1 The connections between three exemplary sustainability criteria of housing and the three pillars of sustainability

	Environment	Economy	Social
Quality	X		X
Affordability		X	X
Differentiation	X	X	X

High quality alone is insufficient unless people can afford to buy (or rent) the products. In other words, affordability (often approximated by relating net income to the average house price) of the dwelling also ought to develop in the same direction and with the same pace as the price level. Plausibly some of the wealthiest residential areas are also economically unsustainable, due to untenable value stability—a narrow part of economic sustainability. To put it simply: a market bubble has occurred in relation to the fundamentals (see Kauko 2010).

The diversity of the product, i.e. types of housing, is crucial in this context. Even if quality and affordability criteria are fulfilled, real estate value stability trends may not be sustainable (even in the economic sense) unless a wide enough range (i.e. product variety generated for most apt selections to be made) of different quality and affordability levels is available on the market. This is because drivers of sustainability such as production technology, community governance and consumption fashions tend to change rapidly and then it is vital not to have neglected any specific housing package even if it may seem marginal at some stage. Or, stated differently, a failure to recognize a potential market trendsetter or other innovation in terms of quality or affordability is bound to have harmful impacts on the evolution of the property portfolio when a steady, moderate price increase in relation to quality and affordability is preferred.

These three sustainability criteria may be related to the three commonly distinguished pillars of sustainability. Table 1 shows these relations. The first two criteria, quality and affordability, are easily accommodated within mainstream economics, i.e., hedonic price models and a welfare economics framework. But this does not apply to the third of the criteria: Analyzing the differentiation of the product range calls for an evolutionary perspective of economic development with a central role of knowledge, diversity, and selection.

The differentiation of real estate products by, e.g., innovation plays an important role in the development of urban sustainability. Some innovations are perceived as positive in sustainability terms, such as green buildings (Lützkendorf and Lorenz 2006; Lorenz et al. 2007) and renewal areas (Hemphill et al. 2004a, b). The integration of ICT into real estate is also worth noting here (Dixon et al. 2005); in fact, the Australian Green Star sustainability rating tool is explicit about incorporating an innovation category: without sufficient documentation of innovative solutions for the development and management of buildings and plots these are not rated as sustainable (Warren-Myers and Reed 2010, p. 204). Other innovations in turn are perceived as negative, such as the products of financial engineering that have led to the American sub-prime crisis. A third category of innovations is ambiguously perceived, such as the application of valuation automata (Mooya 2011) or the establishment of real estate investment trusts (REITs; a vehicle to encourage increased indirect real estate investment; it originates in the USA and was copied in other countries).

2.5 A Strategy for Sustainability Analysis of Real Estate Using Evolutionary Economics

The entanglement between urban real estate markets—in which governmental regulations can be seen as a secondary influence after investment activities—and real estate sustainability can be illustrated with the influence of sustainability certification on investment risks. Certified real estate is likely to benefit not only the tenant but also the landlord via reduction in operating costs, improved image, and security of rent assuming a market situation where the occupants vote with their feet in search of optimal packages of costs and benefits. Thus, sustainable real estate reduces the owner's risk of loss of rent (cf. Lorenz et al. 2008; Eichholz et al. 2009; Fuerst and McAllister 2011; Warren-Myers 2011).

Following these considerations it is, hence, purported that evolutionary economics could be helpful in analyzing sustainable urban real estate. The research strategy brought forward here in order to integrate evolutionary economics and urban sustainability is to analyze the quantitative market data and the qualitative sustainability observations separately at first and then to triangulate these analyses. Further, the suggestion is to use an evolutionary economics approach for the analysis of the quantitative data in order to account for sustainability criteria such as differentiation (see Table 1) that cannot be treated using NCE. It is suggested that the qualitative analysis of the sustainability of real estate be carried out empirically by means of methods such as direct field observation, interviews and focus groups, considering the sustainability of the actual housing and its neighborhood. In the particular contribution documented below, in order to demonstrate the feasibility of the suggested research strategy and to sketch the further research agenda, the approach chosen for the analysis of quantitative data is the SOM, and complementary qualitative data is gathered through expert interviews and site visits.

3 An Exemplary Implementation of the Research Strategy

3.1 The Quasi-Dynamic Modeling Approach Based on the SOM

The SOM, a neural network technique invented by Teuvo Kohonen in 1982, is best defined as a mapping from a high-dimensional data space onto a (usually) two-dimensional lattice of points. This way disordered information is profiled into visual patterns, forming a landscape of the phenomenon described by the data set. In the case of housing data such a landscape looks like a mountain landscape and the phenomenon described is ready to be analyzed in terms of peaks and valleys, that is to say high or low intensities on a given input variable.

The SOM produces a so-called feature map of nodes. This is more formally to be treated as a clustering technique based on iterative runs. The nodes of the feature map represent a characteristic combination of attribute levels; in the case of house prices this characteristic combination of attribute levels tells us about the variation in prices and other features defined through the variables used and their combinations. Such a map, of course, is not a representation of the landscape, but rather a visualization of patterns and clusters inherent in the dataset. The SOM can be used as a tool for the reduction of dimensions of input data, and subsequently for clustering of the observations based on these reduced dimensions. It takes input data, e.g., house prices and, by virtue of the neural network algorithms, clusters it. These clusters can be used to structure the qualitative research, trying to reconcile the pattern of the quantitative analysis with the actual sustainability of the real estate.

The neural network algorithms of the SOM approach require an initial training, which is directed by the datasets and external manipulations of the run (length of the run, shape of the surface, sensitivity of the training, etc.). In the training procedure of the algorithm, the matching between output (response) and input (sample) is usually determined by the smallest Euclidean distance between observation and response. In the case of using data on house prices, each observation represents a market price level for a given location and each response represents a cluster of similar observations that are averaged for that node (i.e. part of the surface). The basic methodological principles are illustrated in Figs. 1 and 2.

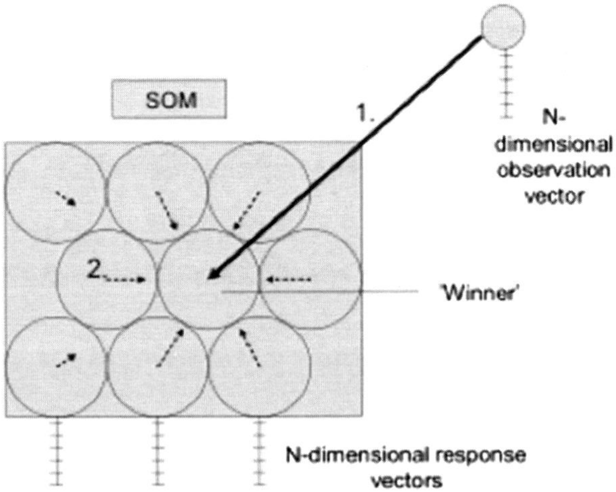

Fig. 1 The two-stage training process in the SOM: 1. Determine the 'winner' node and adjust its weights towards the weights of the observation; 2. Similarly, adjust the weights of adjacent nodes, but less the further they are situated from the 'winner'

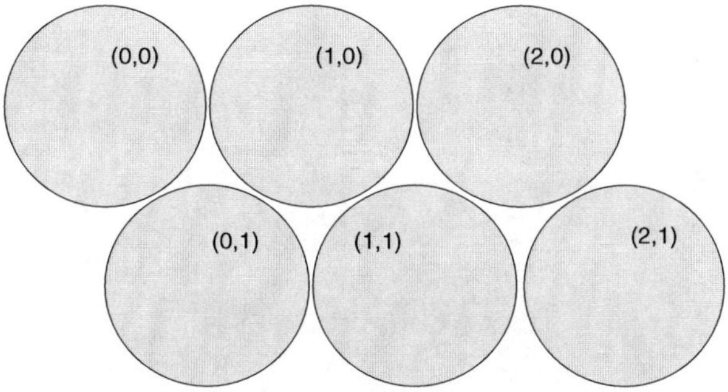

Fig. 2 The situation of the nodes in a three-by-two (3 × 2) map with hexagonal shape

This technique works in three steps: projection, clustering, and discrimination. First, the surface is predefined in terms of the number of potential clusters (nodes) and the parameters for adjustment of this map-like surface (feature map). Second, the map is trained using a data set of m observations (cases) measured as n variables (map layers). Third, the feature map is examined in terms of similarities between nodes and intensities of any nodes for a given map layer. The similarity and intensity of any nodes can also be identified across all map layers when the location of a given node is fixed across these layers by definition—this output is now perceived as a landscape of the data set for a given cross section (see Kauko, 2006a, 2007 and 2009 for further descriptions of this technique).

In a case study carried out by the same author, the data comprise streets in Budapest, for which mean prices and volumes are calculated for four different property types (see Fig. 3). This data is furthermore labeled on the basis of street and district. When examining the 'hits' of the nodes, it is to be observed that a given node is labeled based on one particular case drawn from several similar cases. In this way, the number of 'hits' then also becomes a description of the reliability of the generalization in terms of the measured variables. Thus, if the node has only one hit and given label, the reliability of the node to actually represent exclusively this particular street/district is 1/1 (100 %). If the node has several hits, the reliability declines asymptotically 1/N: two observations give half reliability, three one-third and so forth. For example, in this case study, the SOM generated map layers revealing upper market cases for ten subsequent data cross-sections in two Hungarian cities (Budapest and Szeged).

Thus, the SOM is a neural network technique that allows for the clustering and non-linearity of the dataset. Clustering becomes necessary in case of applications that require either classification or pattern recognition based on data fed into the systems of algorithms which constitute the SOM. Non-linearity in datasets on house-prices occurs due to microeconomic interferences, for example, those of the nearby land uses and amenities, macroeconomic fluctuations and also—although this is debated—from non-economic influences such as behavioral and institutional factors (i.e. those that are accounted for in heterodox economics but not in NCE). Location-specific observations of house prices will be analyzed with the SOM and related to the nature of their market appreciation features.

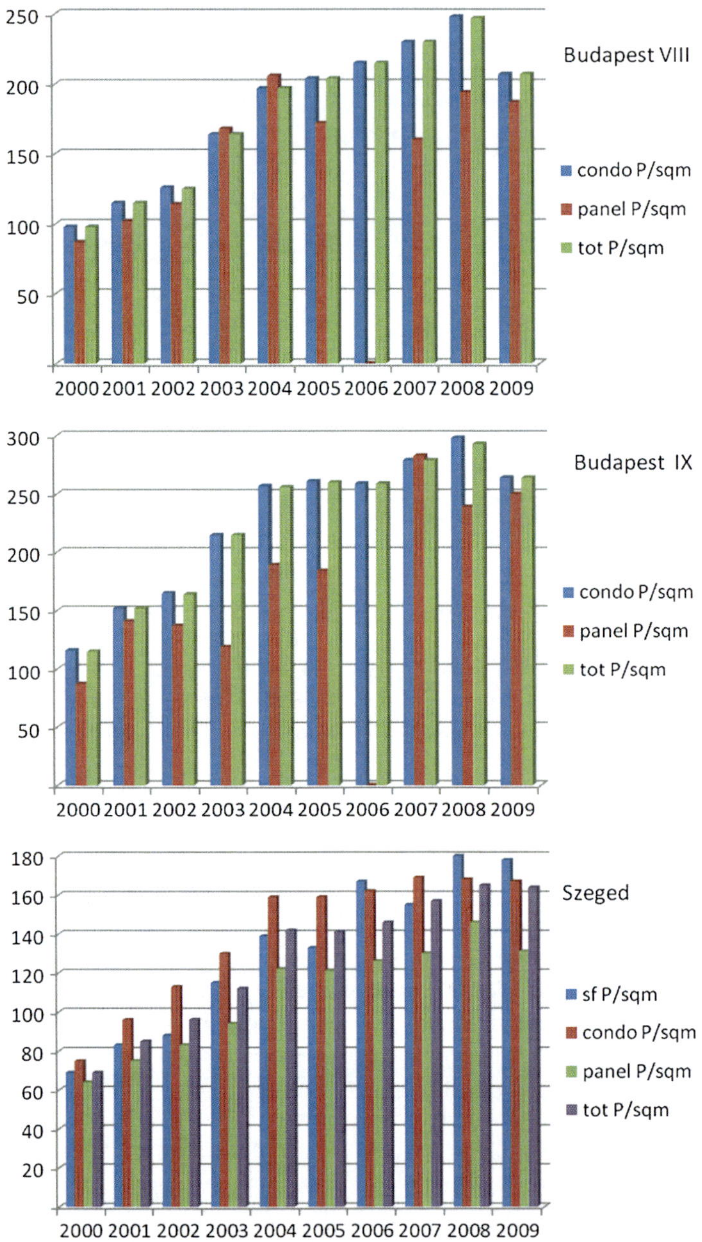

Fig. 3 Per sq.m. house prices ('000 HUF) for the period 2000–2009 for the three property types and total figures for two Budapest districts and Szeged

It should be noted that the SOM approach is merely chosen as an example to implement an evolutionary approach. Many other tools are, potentially, suitable to conduct evolutionary real estate economics analysis. It is unlikely that this choice will lead to a specific methodology. For example, an alternative approach could try to explain the interrelation of house prices and sustainability through variety generation, selection of the fittest, and convergence of the variety generated towards the fittest trajectory—core evolutionary principles as suggested in the seminal contribution of Nelson and Winter (1982) or more recent thinkers (see, e.g., Essletzbichler and Rigby 2007).

3.2 Qualitative Fieldwork Methodology to Study Neighborhood Sustainability

It is to be noted that the SOM is only able to deal with numerical data. However, real estate analysis is about the behavior of actors such as developers and investors as much as about the market outcome. Thus, a feasible method may apply, besides real estate transaction price data, expert interviews about actor motivations and tendencies. We need to triangulate with qualitative information about the sustainability of real estate developments in the selected locations of the city. The ultimate aim here is to see if a certain price development identified from the map outputs could be related to any kind of sustainable development insofar as upgrading has taken place during and before the years of data examination. This will furthermore enable us to try to relate any particular sustainable and unsustainable development elements found in these circumstances with a loosely framed evolutionary theory perspective for real estate economics. Here, it is postulated that the generated diversity of routines (i.e. repeated behavior that is spontaneous, learned, and internalized in short-cuts for problem solving) and subsequent selection of the fittest is the way forward when updating the body of knowledge that governs urban real estate analysis.

The interviews and the site visits are meant to be applied in a fairly pragmatic fashion. For example, in the case study carried out by the same author in Budapest and Szeged, the interviewed experts represented private investors, developers, builders and consultants as well as local government, NGO and academia as these are the sectors which are believed to have comprehensive, yet different, information of the local real estate situations. The questions concerned the motivations for selecting certain business strategies, achievability of different sustainability dimensions, expectations about property/land value increases, compatibility with the surrounding land use and neighborhood characteristics, whether any marketing incentives were considered for the buyers and to what extent any cooperation with the municipality occurred. Here, it needs to be noted that certain gate keepers (who were respondents of the earlier interviews) managed to arrange access to the private developer respondents. The site visits were carried out in a straightforward way: the cases that showed up on the feature maps as upper market cases were visited (to the extent that they were reachable by public transport or walking).

In this case, the qualitative interviews revealed four kinds of sustainability issues in the Hungarian context: First, there is no price-reduction for the lack of sustainable elements in the residential sector; second, the prices are set by the seller in a buyer's market, which is unsustainable by definition; third, corruption of government planning officials leads to escalating land prices; and the current market situation where banks charge interest costs on the sites but the developer cannot realize the buildings (see Kauko 2013b).

3.3 Methodological Learnings from the Case Study

The approach of the SOM and quantitative studies was suggested in order to analyze the extent to which patterns of house prices and sustainability features of the built environment overlap. Depending on the selection of data/variables, that is to say, how many observations and along what kind of dimensions they are defined and measured, as well as the reliability and validity of the data, one is able to illustrate differences in urban structure (including those defined by house price levels across locations and market segments) using the SOM. However, while it is relatively easy to illustrate price premiums, it is far more difficult to relate them to any sustainability factors such as cultural heritage, public transport accessibility, walkability, green areas, mixed stock, energy savings, or maintenance of communal areas. This is because such qualitative info cannot be processed with the SOM. Therefore, field inspection is needed to reveal the plausible reasons for a price premium, and whether this can be analyzed further according to sustainability criteria. Lastly, a caveat is in order: As the SOM is about individual-level processes and spontaneity in the formation of patterns, it is possible that complexity theory is better equipped for providing theoretical guidance to this research idea than what EEG is. This will be a new research direction (see Foxon et al. 2013).

As already noted, the SOM is only one of many possible tools to implement an EE framework. When contemplating the usefulness of such an extensively numerical and heavily computerized approach, the question is whether any other method could be worth consideration. Obviously the SOM type of approach is not widespread—its heyday never really came, although a peak in the number of promising attempts at applying it occurred during the 1990s (Carlson 1998; Kauko 2006a and 2007). Today, the quantitative research frontier is moving with the progress of machine learning approaches such as cellular automata on one hand (e.g., Meen and Meen 2003), and more actor-oriented approaches such as multi-criteria decision making approaches on the other (e.g., Kauko 2006b). Such approaches have undoubtedly become more popular than the SOM—or other neural network-based approaches—in socioeconomic analysis. Nonetheless, the appeal of the SOM is that its functioning principles and outcomes can be translated into the same language that econometricians use and spatial-visual analysts speak. Given the eclectic nature of this research area, this is no small advantage.

The beauty of the SOM approach lies in its ability to map the variation of the data set, which then gives a basis for further inquiries into special cases using a more qualitative approach. The methodological findings from the two Hungarian case studies clearly confirm this; it is easy to illustrate price variation across the dataset, but difficult to relate the intensities to sustainability criteria—for that interviews and field inspection are suggested. However, this piece of research also pointed to an interesting empirical relation: the cultural heritage buildings were the most sustainable real estate feature in the rather homogeneous Budapest market context whereas its more heterogeneous Szeged counterpart showed all sorts of sustainable features typical of a smaller, more walkable and less polluted town. Based on these findings we can furthermore note a loose fit between sustainable real estate analysis and an evolutionary (or complexity) meta-theory, where findings from the former realm may (or may not) be supported by the latter (see Kauko 2013a, b).

4 Discussion of the Suggested Research Strategy

4.1 Status Quo of the Suggested Research Strategy

When setting a novel research agenda based on the learnings presented above, the present research strategy merges several traditions that are 'evolutionary' (and complex) in the broad sense insofar as these share a key characteristic and aim towards a common frame. First, the aspatial tradition of evolutionary economics (1980s); second, the spatial tradition (i.e. the evolutionary economic geography traditions championed by Boschma, Martin and others since the 1990s); third, the sustainable development tradition (2000s).

It is worth noting that when evolutionary theorization began in the early 1980s, the sustainable development discourse was unknown; for instance, Nelson and Winter firmly discuss the strategies of economic agents in the context of normal profit-making only. Since then, however, the sustainability issue has forcefully been brought to the fore. At present, even mainstream real estate economists point out that investment informed by sustainability criteria, such as green buildings, can have direct or indirect economic benefits for market-based actors, or non-economic benefits (ethical behavior) for actors with soft budget constraints (see Eichholz et al. 2009). Examples of direct economic benefits include long-term cost savings and increased employee productivity, and examples of indirect economic benefits include increased reputation, a more loyal workforce, and minimized risk with regard to a future tightening of regulations. Thus, achieving these goals leads to various benefits for the business occupying the facilities and eventually to these benefits being capitalized in higher property values.

The question posed here is whether the core evolutionary principles of variety generation, selection of the fittest, and convergence of the variety generated towards the fittest trajectory, for example, in the sense suggested by the seminal contribution of Nelson and Winter (1982) or more recent thinkers (see, e.g., Essletzbichler and Rigby 2007), can be applied as a support to results regarding the development of real estate prices using the SOM approach, and the evaluation of urban sustainability using expert interviews and qualitative findings (Kauko 2013b).

To sum up the research strategy, the numerical and visual analysis carried out using the SOM approach is here confronted with corresponding findings regarding urban sustainability. On top of this, a further aim of the study is to examine the interview material to check if there is an association between any kind of innovativeness related to how a more sustainable urban real estate context might be achieved, learning and networks (or the lack thereof) and the price increases (or decreases) that are identifiable from the map.

4.2 Challenges of Furthering the Research Strategy

Many challenges go beyond the specific learnings from the case study and the presented concept of the research strategy. They are drawn from earlier general research and analysis on evolutionary and sustainability approaches. Here, a variety of particular questions come to mind:

- Most evolutionary economic theory deals with firms. Is the evolutionary approach to real estate then also only about firms, not residents? If so, how would developers and investors be accommodated within such an approach?
- Is the evolutionary approach only about regional patterns, not urban ones? If so, how would a metropolitan region fit in?
- Can this be connected at least to meta-theory, even if no strict coupling to evolutionary economic theory is possible (this would be due to the fundamental lumpiness and longevity of the actual real estate products together with the inherent conservatism of many real estate actors). If so, in what way could the evolutionary framework still provide guidance?
- What about the compatibility between the similarity-based clustering of the SOM and the EEG notion of clusters? Is the former only an intuitive term whereas the latter is part of an established paradigm of its own?
- When the EEG notion of entrepreneurship and innovation (following Boschma, Franken and others) was established, the topic was that of the industrial production sectors (see also Rigby and Essletzbichler 2006). Is it at all conceptually feasible to include long-term aspects including social and environmental issues into that body of knowledge?
- EEG postulates the occurrence of lock-ins as a result of dysfunctional practices. Martin (2012) sees lock-in as an evolutionary element of policy or behavior that undermines any logic of decisions being rational. Hassink (2005) in turn

purports that such failures may be unlocked by forming 'learning clusters', given the necessary social capital and other favorable conditions. How much of this conceptualization can be brought onto a sustainable urban real estate context? For example, can such dysfunctional (and by implication unsustainable) situations be corrected by reverting to a business strategy based on niche developers and niche markets? Martin and Sunley (2006) teach us that the mechanisms that lead to lock-in and de-locking processes are either global or local in character. This should fit the real estate market as an increasingly global but essentially local entity.
- What about the connection of evolutionary theory lines to traditional hedonic type price theory that explains price changes in relation to changes in the balance of amenity and nuisance factors? Is sustainability theory here to offer a new paradigm where environmental, social, and economic interests do not conflict in the long run, or is this only the same essential model but with a longer time frame?
- Martin (2012, p. 182) argues that an institutionalist view can be incorporated into the evolutionary view. We accept that the old institutional economics is already found compatible with real estate (Kauko 2013b), does this now imply a compatibility with the evolutionary economic perspective too?

Martin and Sunley (2007) propose complexity-theoretical thinking (when refined) as an alternative to path dependence and EEG. The two lines of theory work on partly different assumptions, but to some extent also share a lot of similar assumptions. Complexity theory has the following differences to EEG:

- The concept of spontaneous order out of self-organization in systems is a broader concept than the concept of selection of the fittest and convergence towards that pattern, which EEG follows.
- Markets and individual behavior are placed in a central role in complexity theory, which means that complexity thinking resembles NCE after all. It can therefore be argued that also EEG ought to focus on these elements rather than to downplay markets and emphasize only the institutional contexts.

On the other hand, the following similarities between the two theoretical approaches can be noted (see Martin and Sunley 2007):

- The shared assumption of economic development being driven by the growth of knowledge.
- The belief that mathematical and computer simulations are justified methodologies to explain or understand the functioning of (complex) systems.

As hinted at above, the possibility exists that the proposed research will fit better with complexity theory than with EEG due to its explicit employment of self-organizing features and micro-level processes. Such a view would definitely put these efforts closer to NCE than what is the case when following other evolutionary lines of enquiry (cf. Renigier-Biłozor 2008). In this vein, the contributions of Ludwig von Bertalanffy (1901–1972) to social science and psychology must be noted; he emphasized systems of symbols in relation to human behavior, which, in his view, distinguished humans from other species (Weckowicz 1989).

5 Conclusions

It looks inevitable that the way we perceive the built environment is undergoing change, regardless of one's disciplinary background or ideological stance. It has to, because of the new requirements of clean and renewable energy together with energy efficiency, attached to the sustainable development agenda since the Rio congress and Local Agenda 1992; more recently, also the financial crisis management agenda is a reality. A third, more traditional strand of this debate concerns how to combat social problems related to poverty and inequality (see Manzi et al. 2010).

The urban real estate economy can be approached from different angles involving evolutionary economics and sustainable development concepts. The value of urban real estate can be seen either via the physical route related to sustainable development, or the asset route relating to economic development. The innovation purported in this paper concerns the way urban and real estate sustainability arguments, for instance, about increased demand orientation, attention to niche products, and government support, can be seen as manifestations of the evolutionary dynamic process which in general is composed of three stages: (1) generation of diversity, using innovative and knowledge-based strategies; (2) selection of the fittest strategy or product; (3) convergence towards that fittest, using imitation. The task discussed in this paper is how to demonstrate this sequence empirically in an urban real estate context. The proposed research strategy attempts to tie together these two lines of enquiry through the sustainability concept and evolutionary theory. Furthermore, a related line, complexity theory, might even be better placed to guide this analysis.

Acknowledgments The paper is based on a presentation held on 12 April 2012 in Vienna, for the EMCSR conference, symposium K. The author is thankful for the comments received from that session.

References

Boschma, R., Frenken, K.: The emerging empirics of evolutionary economic geography. J. Econ. Geogr. **11**, 295–307 (2011)

Carlson, E.: Real estate investment appraisal of land properties using SOM. In: Deboeck, G., Kohonen, T. (eds.) Visual Explorations in Finance with Self-Organizing Maps, pp. 117–127. Springer, New York (1998)

d'Amato, M., Kauko, T.: Sustainability and risk premium estimation in property valuation and assessment of worth. Build. Res. Inf. **40**(2), 174–185 (2012)
Dixon, T., Thompson, B., McAllister, P., Marston, A., Snow, J.: Real Estate and the New Economy. The Impact of Information and Communications Technology. RICS Real Estate Issues. Blackwell Publishing, Oxford (2005)
Dixon, T.: The property development industry and sustainable urban brownfield regeneration in England: an analysis of case studies in Thames Gateway and Greater Manchester. Urban Stud. **44**(12), 2379–2400 (2007)
Dopfer, K.: Kenneth boulding: a founder of evolutionary economics. J. Econ. Issues XXVIII **4**, 1201–1204 (1994)
Eichholz, P., Kok, N., Quigley, J.: Why Do Companies Rent Green?. Real Property and Corporate Social Responsibility. RICS Research, London (2009)
Essletzbichler, J., Rigby, D.L.: Exploring evolutionary economic geographies. J. Econ. Geogr. **7**, 549–571 (2007)
Foxon, T.J., Köhler, J., Michie, J., Oughton, C.: Towards a new complexity economics for sustainability. Camb. J. Econ. **37**(1), 187–208 (2013)
Fuerst, F., McAllister, P.: Green noise or green value? measuring the effects of environmental certification on office values. Real Estate Econ. **39**(1), 45–69 (2011)
Hassink, R.: How to unlock regional economies from path dependency? from learning region to learning cluster. Eur. Plann. Stud. **13**(4), 521–535 (2005)
Hemphill, L., Berry, J., McGreal, S.: An indicator-based approach to measuring sustainable urban regeneration performance: part 1, conceptual foundations and methodological framework. Urban Stud. **41**, 725–755 (2004a)
Hemphill, L., Berry, J., McGreal, S.: An indicator-based approach to measuring sustainable urban regeneration performance: part 2, empirical evaluation and case-study analysis. Urban Stud. **41**, 757–772 (2004b)
Kauko, T.: Urban Housing Patterns in a Tide of Change: Spatial Structure and Residential Property Values in Budapest in a Comparative Perspective. DUP Science Publication, Delft (2006a)
Kauko, T.: What makes a location attractive for the housing consumer? preliminary findings from metropolitan Helsinki and Randstad Holland using the analytical hierarchy process. J. Housing Built Environ. **21**(2), 159–176 (2006b)
Kauko, T.: A heterodox economic analysis of the housing market structure in Budapest using neural network classification. J. Real Estate Lit. **15**(1), 85–124 (2007)
Kauko, T.: The housing market dynamics of two Budapest neighborhoods. Hous. Stud. **24**(5), 587–610 (2009)
Kauko, T.: Value stability in local real estate markets. Int. J. Strateg. Property Manage. **14**, 191–199 (2010)
Kauko, T.: An evaluation of the sustainability of inner city residential projects. Hous. Theory Soc. **28**(2), 144–165 (2011)
Kauko, T.: Self-Organizing Map Algorithms to Identify Sustainable Neighbourhoods with an Example of Szeged (Hungary). GIS Ostrava 2013—Geoinformatics for City Transformation, January 21–23 (2013a)
Kauko, T.: A Demonstration of Sustainability Arguments Using House Price Data. In: 20th Annual Conference of the European Real Estate Society, pp. 146 (2013b)
Lorenz, D., Trück, S., Lützkendorf, T.: Exploring the relationship between the sustainability of construction and market value: theoretical basics and initial empirical results from the residential property sector. Property Manage. **25**(2), 119–149 (2007)
Lorenz, D. and other members of the RICS Advisory group on Sustainable Property Investment and Management: Sustainable Property Investment and Management—Key Issues and Major Challenges, RICS (2008)
Lützkendorf, Th, Lorenz, D.: Using an integrated performance approach in building assessment tools. Build. Res. Inf. **34**(4), 334–356 (2006)

Macintosh, A.: Visions of Tomorrow to Plan our Lives Today—Sustainable Development and the Real Estate Industry. EU Sustainable Energy Week, Brussels (2010)

Maclennan, D., Tu, Y.: Economic perspectives on the structure of local housing systems. Housing Stud. **11**(3), 387–406 (1996)

Manzi, T., Lucas, K., Lloyd-Jones, T., Allen, J.: Understanding social sustainability: key concepts and developments in theory and practice. In: Manzi, T., Lucas, K., Lloyd-Jones, T., Allen, J. (eds.) Social Sustainability in Urban Areas. Communities, Connectivity and the Urban Fabric, pp. 1–28. Earthscan, London (2010)

Martin, R.: The local geographies of the financial crisis: from the housing bubble to economic recession and beyond. J. Econ. Geogr. **11**(4), 587–618 (2011)

Martin, R.: (Re)Placing path dependence: a response to the debate. Int. J. Urban Reg. Res. **36**(1), 179–192 (2012)

Martin, R., Sunley, P.: Path dependence and regional economic evolution. J. Econ. Geogr. **6**(4), 395–437 (2006)

Martin, R., Sunley, P.: Complexity thinking and evolutionary economic geography. J. Econ. Geogr. **7**(5), 573–601 (2007)

Meen, D., Meen, G.: Social behaviour as a basis for modelling the urban housing market: a review. Urban Stud. **40**(5–6), 917–935 (2003)

Mooya, M.: Of mice and men: automated valuation models and the valuation profession. Urban Stud. **48**(11), 2265–2281 (2011)

Nelson, R.R., Winter, S.G.: An Evolutionary Theory of Economic Change. Belknap Press of Harvard University Press, Cambridge (1982)

Nelson, R.R., Winter, S.G.: Evolutionary theorizing in economics. J. Econ. Perspect. **16**(2), 23–46 (2002)

Plummer, P.: Economic geography, by the numbers? In: Tickell, A., Sheppard, E., Peck, J., Barnes, T. (eds.) Politics and Practice in Economic Geography, pp. 199–209. Sage, London (2007)

Ratcliffe, J., McIntosh, A., Brown, S.: Built Environment Foresight 2030: The Sustainable Development Imperative. The University of Salford, Futures Academy and King Sturge, RICS Foundation (2010)

Renigier-Biłozor, M.: Residuals analysis for constructing 'more real' property value. In: Kauko, T., d'Amato, M. (eds.) Mass Appraisal Methods: An International Perspective for Property Valuers, pp. 148–163. Blackwells, Oxford (2008)

Rigby, D.L., Essletzbichler, J.: Technological variety, technological change and a geography of production techniques. J. Econ. Geogr. **6**(1), 45–70 (2006)

Robson, A.J.: Evolution and human nature. J. Econ. Perspect. **16**(2), 89–106 (2002)

Warren-Myers, G.: Valuing sustainability: a new challenge for the profession. Australia and New Zealand Property J. **3**(3), 157–164 (2011)

Warren-Myers, G., Reed, R.: The challenges of identifying and examining links between sustainability and value: evidence from Australia and New Zealand. J. Sustain. Real Estate **2**(1), 201–220 (2010)

Weckowicz, T.E.: Ludwig von Bertalanffy (1901–1972): A Pioneer of General Systems Theory. CSR Working Paper No. 89.2, Center for Systems Research, University of Alberta, Edmonton (1989)

Organized and Disorganized Complexities and Socio-Economic Implications in the Northern Ruhr Area

Hans-Werner Wehling

Abstract The present-day Ruhr region is characterized by a general south-north (regional) divide in terms of urban and social structures. The chapter tries to show that the multi-phase nineteenth and early twentieth century industrialization has caused different urban systems of (dis)organized complexity; that the multi-phase process of post-war de-industrialization has been disrupting the urban system on different scales; that, within the region, different urban systems came into being and were changed by the decisions/interests of influential groups in various alliances; that two multi-phase periods of immigration contributed to these different urban systems; and that current planning approaches are increasingly adjusting themselves to the forms of urban complexity.

Keywords Industrialization · De-industrialization · Segregation · Organized complexity · Disorganized complexity

Using Weaver's (1948) fundamental concept, Jane Jacobs was among the first who, in the famous last chapter of her *The Death and Life of Great American Cities* (Jacobs 1961), referred to the city as "organized complexity" which, unlike "disorganized complexity", cannot be dealt with by means of probability techniques or statistical mechanics. In contrast to physical sciences but similar to many complex natural systems, cities are self-organizing systems of organized complexity and should be addressed as a living organism by thinking about processes (cf. Mehaffy 2004).

More than 40 years after Jacobs, and summarizing the wide scope of approaches launched ever since, Michael Batty states:

> Cities are no longer regarded as being disordered systems. Beneath the apparent chaos and diversity of physical form, there is a strong order and a pattern that emerges from the myriad of decisions and processes required for a city to develop and expand physically. Cities are the example par excellence of complex systems: emergent, far from equilibrium,

H.-W. Wehling (✉)
University of Duisburg-Essen, Institute of Geography, Schützenbahn 70, SR 107, 45127 Essen, Germany
e-mail: hans-werner.wehling@uni-due.de

C. Walloth et al. (eds.), *Understanding Complex Urban Systems: Multidisciplinary Approaches to Modeling*, Understanding Complex Systems,
DOI: 10.1007/978-3-319-02996-2_6,

> requiring enormous energies to maintain themselves, displaying patterns of inequality spawned through agglomeration and intense competition for space, and saturated flow systems that use capacity in what appear to be barely sustainable but paradoxically resilient networks. (Batty 2008, p. 769)

Batty not only regarded dynamics as crucial to complex systems but also stressed the importance of qualitative changes in the form of phase transitions, arguing that

> complex systems have 'tipping points' where unusual sets of conditions come together and fire the system in one way or another ... phase transitions are also associated with qualitative changes such as that generated often endogenously within the system such as the development of disruptive technologies or dramatic switches in human behavior and preferences. (Batty 2007, p.12)

The idea of phase transitions generated by changes in human behavior makes the concept of complexity interesting to the social sciences and, in particular, to studying urban systems:

> Cities are regarded as organizational, social structures that are changed through decision-making of various kinds and it is this nexus of decisions across all scales and through all times that forms the web of complexity that is basic to the social sciences. (Batty 2007, p. 26)

Facing the fact that the present-day Ruhr region is characterized by a general south-north (regional) divide in terms of urban and social structures, the following chapter tries to show

- that the multi-phase process of regional industrialization has caused different urban systems of (dis)organized complexity,
- that the multi-phase process of de-industrialization must be seen as a process that has been disrupting the urban system on different scales,
- that the different urban systems came into being and were changed by the decisions/interests of influential groups in various alliances,
- that two multi-phase periods of immigration contributed to these different urban systems. The social patterns of the northern Ruhr area (Emscher zone) will be analyzed at the beginnings of the twentieth and twenty-first centuries,
- that current planning approaches are attempting to adjust themselves to these forms of urban complexity.

1 Disorganized Complexity: Phases of Industrialization, Immigration and Segregation

1.1 Development of Industrialization in the Ruhr Area Until World War II

Since the mid-nineteenth century, coal mining and the iron and steel industries became the basis of the economic development of the Ruhr. The expansion of the industries and their technological development were fostered by a series of self-intensifying cause-symptom processes that resemble the Kondratieff cycles of basic technologies: the transfer of expertise and metallurgical innovations allowed a greater and wider output to be achieved more cheaply and more efficiently; innovations of the international finance system gave birth to new types of enterprise and thus increased the number of new companies; shipping, canals and particularly the railway set a third flywheel in motion, as they not only consumed coal and steel but also opened up new markets; since 1871, Germany's unification caused a high demand for infrastructure for trade, armed forces and administration; the new state became both investor and client, thus guaranteeing permanent orders and increasing sales to the regional customers. Structurally, the Ruhr economy developed along the product lines of iron and particularly of coal and achieved its highest diversity in the 1930s.

Spatially, the basic industries exhibited different demands and patterns. The labor force and market-orientated iron and steel industry preferred sites in the vicinity of the existing cities as they were better connected to the existing traffic systems. In contrast, coal mining had to adjust itself to the geological conditions. In the southern Ruhr valley, the coal measures crop out and had been exploited by gallery mining in pre-industrial times already. But as the seams in the north dive down to more than 1,000 m, it was not before the technologies of shaft mining and a controlled draining by pumps had been developed that deeper and more valuable coal seams could be tapped. Thus, the carboniferous geology caused a northward movement of coal mining; as this industry was essential to all other industries including iron and steel, paramount differences of the industrial structures ensued. Zones of industrial development came into being that differed in terms of technical standards, production methods, productivity, dimension of sites, and impact on the landscape. From 1843 onwards the shaft mining technique allowed coal mining to spread from the Ruhr valley into the Hellweg zone where some small cities (such as Duisburg, Essen, Dortmund) along the medieval trading route of the Hellweg had dominated the pre-industrial economy. Since the late 1860s and particularly fueled by the demands of the new German Empire, coal mines were also spreading farther north into the boggy areas of the Emscher valley that had deterred any settlement so far and that were still affected by annual flooding and epidemic diseases.

As far as decision-making is concerned, three groups with different interests and therefore different impacts on the systems can be distinguished; these are the industrialists/industrial companies, the mayors/city councils of the existing/developing cities, and the political authorities (Prussia, German national government, North-Rhine Westphalian state government). In the course of the regional development these three groups formed various alliances. The industrialists and the political authorities formed a close alliance which, in the first phases, widely disregarded the interests of the cities but proclaimed industrial demands as predominant in order to develop an industrial area of national importance.

From 1852 to 1871 the regional population increased by 46.8 % from 374,400 to 703,100 (Steinberg 1985, p. 22), mainly focused on the Hellweg zone. The demand for workers could still be satisfied either on the local labor markets or by miners and iron-workers migrating in from other parts of the province of Westphalia and the Rhineland; a combination of industrial and agricultural work was still common. So in the early phases of industrial development, the industrial workers who migrated into the Ruhr valley and into the Hellweg zone did not remarkably differ from the local population in terms of culture, religion and life-styles and thus did not substantially change the region's social structures.

1.2 Formation of Zones of Different Ethnicities

With the economic consequences and impulses ensuing from the foundation of the German Empire, however, the super-sized coal mines and iron and steel works of the Emscher valley were not only confronted with an enormous per se demand for labor force but also had to cope with their locations in the nearly unpopulated areas along the river Emscher. Therefore, the regional industrialists sent out agents into the east Prussian provinces and to Poland in order to recruit young rural workers and their families to work in the large works developing in the Emscher valley. In the course of three waves of immigrations from 1871 to World War I, some 700,000 people (Steinberg 1985, p. 85) were settled in the areas on both sides of the river Emscher.

Since c. 1900 the dynamic self-intensifying cause-symptom circles mentioned above were condensed into an (urban-)industrial complex that had pervaded all fields of the regional society by the beginning of the twentieth century. This is particularly true for the areas on both sides of the river Emscher, where larger production sites and an urban fabric had come into being that had only been laid out according to the industrial needs. Workers' life-styles left their marks on housing and living, and an economic middle-class or even a cultural bourgeoisie was largely missing.

In social terms, the differences between the Hellweg zone and the Emscher zone had often been referred to only on a zonal level (Fig. 1). Many contemporaneous writers, such as, e.g. distinctly Spethmann (2011, pp. 1427ff), addressed the Hellweg zone as that part of the Ruhr that had decently, in a socially balanced way, and in an appropriate urban form been transferred into the industrial age, whereas the Emscher zone—even if gazed at for the technical and economic achievements—was xenophobically claimed to be an "unfinished" and "untamed" area of unwanted but needed population that was accustomed to different life-styles, different ideas, and sometimes even different languages. Such sweeping objections often failed to recognize that this piece of the dawning new working-class mass society exhibited a remarkable structural and spatial diversity as is evident in the facts

- that the entrepreneurs tried to achieve a homogenous labor force and in doing so also paid attention to a more or less consistent denomination of their workers. Catholic entrepreneurs therefore preferred a recruitment of workers in catholic regions like Poland and Upper Silesia, whereas protestant industrialists had their

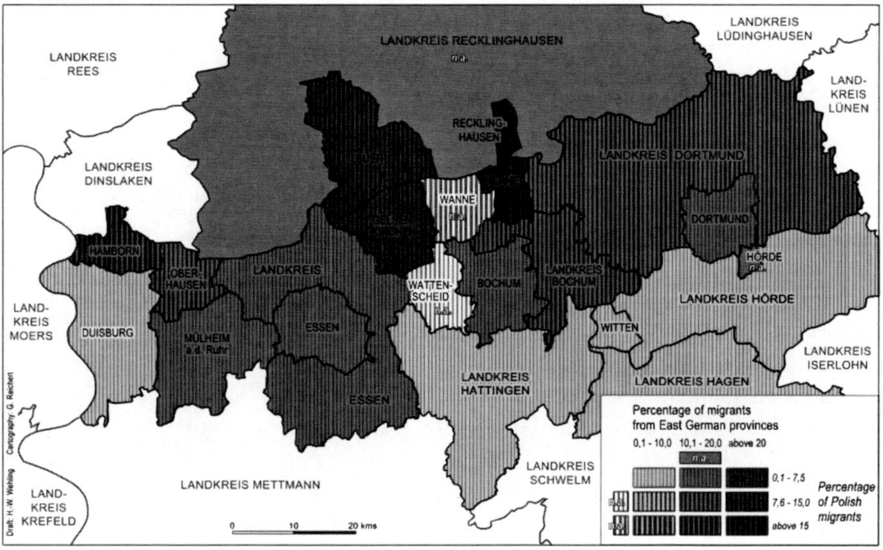

Fig. 1 Migrants from Poland and the East Prussian provinces in the urban and rural districts of the Ruhr area, c. 1910 (*Data source* Steinberg 1985, p. 90)

recruitments carried out in the provinces of Posen, West Prussia as well as in East Prussia and its region of Masuria;

- that the immigrants exhibited 'classical' patterns of residential segregation; as far as the immigrants from East Prussia are concerned, Franke (1936, p. 40) has shown that citizens from Allenstein and Ortelsburg mainly settled in Gelsenkirchen, those from Neidenburg and Soldau mainly in Wattenscheid, and that people from Lötzen migrated to Wanne and those from Osterode to Bochum; the city of Gelsenkirchen became the Masurian gateway into the Ruhr area;
- that the migrants from Masuria and Poland were both different in terms of language. The Poles ("Ruhrpolen")—with highest proportions in the urban districts of Wanne (26.8 %), Recklinghausen (23.1 %), Herne (21.6 %) and Hamborn (17.4 %)—were avoided by the Germans inside and outside the Emscher zone. The strict Catholicism unknown so far in the Ruhr area and the cultural and political relations which the Poles maintained to their places of origin ended up in fierce controversies between nationalistic German politicians and the leaders of the Polish movement at the dawn of World War I. Nevertheless, Polish workers were appreciated for their working morale.

1.3 Increased Differences Between Urban Centers and Industrial-Suburban Patterns

In the early 1930s, the regional industries achieved a technical maximum in terms of efficiency and paramount connectivity. The overwhelming predominance of industrial demands and the decision-making influence of the alliance of industrialists and national political authorities ceased, particularly in the Hellweg zone cities. And it was there that the influence of the city councils brought forward economic and urban systems beyond the mere industrial needs. Thus the economic and functional differences between the Hellweg and the Emscher zone increased in the 1930s with the majority of the Hellweg cities turning into regional service centers. In the incoherent industrial/suburban Emscher zone land-use patterns, however, nearly any regular urban development was restricted if it was to thwart industrial interests.

On the eve of World War II, the Ruhr area exhibited an overall industrial system of disorganized complexity: The interests and decisions of both the industrialists and the national (economic) political authorities had prevented self-organization and any organic development of urban systems in the Ruhr area, specifically in the Emscher zone. All developments were adjusted to the requirements of the industrial development and the decisions of the entrepreneurs, leading to zonal, but not layered, differences. For instance, various forms of immigration increased the interzonal differences and to a certain degree also the intrazonal ones.

Both the interzonal differences and the industry-orientated Emscher zone's spatial patterns were reproduced and reinforced in the course of the *wirtschaftswunder* years of the 1950s and 1960s. The 1910 diversity of the social structure, however, could still be traced but had widely been swept away by the euphoric mood of those years. The economic progress achieved in this region by people from nearly all parts of Germany and the fact that many Polish migrants had given up their native language and, partly forced by official campaigns, had their names Germanized, helped constitute the myth of the Ruhr as a social melting pot. This myth reached out into the 1950s and 1960s, but sometimes camouflaged the still existing prejudices, the stories of frequent job changes, dismissals, and failed assimilation and integration.

2 Organized Complexity: Phases of De-Industrialisation and Social Diversification

The war economy and the post-war reconstruction phase gave additional impulses without which the coal mining and steel industries would have lost their dominant influence on the regional economy already before. Retrospectively, it might be argued that already the 1930s mark the beginning of a phase transition according to Batty, but this was only in parts effective in the Hellweg zone. War economy and the *wirtschaftswunder* years widely prevented a new phase becoming effective all over the Ruhr area.

2.1 Developments of De-Industrialization in the Ruhr Area After World War II

The crisis of the regional coal industry that started in the late 1950s was due to the increasingly costly methods of mining, to a liberalized international trade that allowed cheaper coal to be imported from abroad, and to the fact that coal was increasingly substituted by oil and gas on the energy market. Since the 1970s the regional iron and steel industry had to face new competitors on the international market, which eventually concentrated the ordinary steel production in emerging countries like China, India, and Brazil, or on the Mediterranean coasts and confined the production of highly innovative special steels to some sites on the 'Rhine front' in Duisburg.

Spatially, the process of de-industrialization more or less followed the same direction as the historical industrialization. The closing down of mines started in the Ruhr valley and in the Hellweg zone and, since the early 1980s, also encompassed the mines in the Emscher zone. The once many hundreds have been reduced to only three mines in 2012, and the last mine is due to shut down in 2018. Public subsidies allowed the coal mining industry to retain or 'smoothly' to melt off hundreds of thousands of jobs. Protected by a coalition of interests (entrepreneurs, workers/trade unions, and the national/regional political bodies) the conversion of the 'old industrial society' into a 'new service-based society', and finally into a knowledge-based society could therefore be arranged in quite a socially consistent way, which has worldwide been accepted as a successful model.

2.2 Increasing Structural Differences Between Hellweg and Emscher Zones

As de-industrialization started in the Ruhr valley and in the Hellweg zone, it was in these zones as well that the first programs for economic and urban change were launched. By about 1980 the majority of the urban development schemes, funded by the national and regional governments, had been realized in the Hellweg zone; new and modern housing quarters had been built, four universities had been established and 58 % of the regional jobs in service industries were concentrated in the Hellweg cities. In the Emscher zone, however, the super-sized mines and iron and steel works had been able to overcome the first decades of crises by methods of rationalization. Their still lasting influence on the local property market impeded nearly any form of urban development. However, when the regional crises reached this zone, the overall economic development had deteriorated remarkably and the low local and national budgets restricted programs of economic and urban change.

2.3 Layers of Segregation Leading to Organized Complexity

These crises, which caused an increasing number of job losses in the basic industries, launched three demographic and socio-economic processes.

- To escape from unemployment, younger, well-educated, and active singles and families left the region in the 1970s and early 1980s; up to the late 1990s, this out-migration was accompanied by a core-periphery migration of those younger households that retained their working-places in the core cities but wanted to live in the green suburban peripheries. Both migration processes caused a concentration of left-behind older and aging population in the core cities that consequently have earlier been affected by the current demographic change than the districts on the periphery (*demographic segregation*).
- Up to the 1980s the Emscher zone remained the production centre of the two basic industries and attracted laborers from both outside and from the southern parts of the region. Unlike the Hellweg zone, where the way had been eased for structural change, in the Emscher zone the old industrial structures remained and were strengthened by these migration flows from the 1980s onwards. As each coal mine or steel work had been the center of miners' settlements and thus each shut-down took the 'heart' out of the (sub)urban fabric, a pattern of holes developed and fragmentation increased; 'problem households' (old, poor, unemployed, migrants, single) remained or moved in. With the crises progressing and unemployment increasing, the old-industrial working-class structure of the Emscher zone became more and more decoupled in terms of economic chances, education, income, and public participation (*social segregation*).
- Since the 1973 energy crisis, the recruitment of labor force from the Mediterranean, particularly from Turkey, has been stopped. But as the foreign workers still living in the Ruhr were allowed to let their families follow, the foreign population kept on increasing. Since 1990, the ceasing immigration from the Mediterranean was replaced by migrants from East European countries. Most of the migrant households still show a generative attitude different from German households. Higher birth rates are to increase their share of population and thereby help partly to compensate the overall population decrease caused by the aging of the German population. Depending on the structural and spatial development of the heavy industries, the majority of the workers once recruited from abroad followed the shift of the production focus in the course of the deindustrialization process. Since the late 1970s, foreign (mostly Turkish) households as well as nationalized Turkish households (Germans with a migration history) of the first, second, and sometimes even the third generation concentrate in the Emscher zone. Here, they have become part of the labor force; here, they received their share of unemployment, and it is here that self-contained Turkish communities came into being that dominate the local social environment (*ethnic segregation*).

In the Emscher zone, various forms of segregation overlap in many ways and produce a complex pattern of multiply deprived local districts. The number of households living on public subsidies is a good indicator for such multiple deprivation which, combined with the shares of foreigners and migrant households, clearly marks the Emscher zone as the severely deprived part of the Ruhr (Fig. 2). Whereas the southern and the central parts of the cities of Duisburg, Essen, and Dortmund are middle-class housing areas that have undergone remarkable processes of urban renewal in the course of the structural change, the northern parts of these cities as well as, e.g., the cities of Oberhausen, Gelsenkirchen and Herne were the working-class areas of the 1980s and early 1990s and are now the areas where the several forms of segregation form complex patterns of multiple deprivation.

The ethnic poverty quarters in the Emscher zone can be seen as the most complex spatial coincidence of effects and processes that ensue from de-industrialization. With a functioning job market, segregated ethnic communities can serve as gateways into the receiving society with persons of the second generation being integrated into that society and leaving these quarters to settle elsewhere. Such has happened in the 1970s and early 1980s. If, however, the job market is limited due to de-industrialization, economic stagnation and transformation, second and third generations of households with a migration history become trapped in these quarters (cf. Kersting et al. 2009, p. 145).

To sum up, the type of complexity of the industrial/urban system changed in the phase of de-industrialization. The system of disorganized complexity, with spatially and socially distinct zones, faded during the post-war decades. In particular,

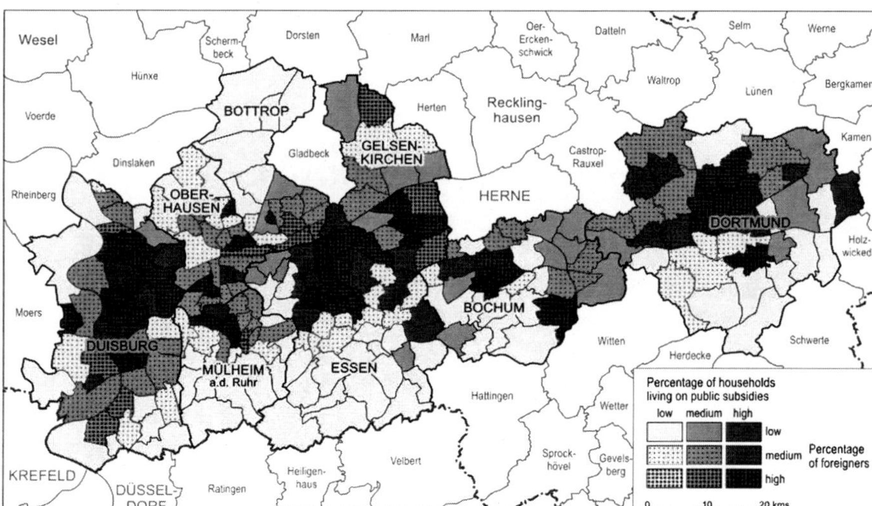

Fig. 2 The social divide of the Ruhr area—multiple layers of social and ethnic segregation, 2008 (*Source* Kersting et al. 2009, p. 143)

various processes of segregation led to a system of (self-)organized complexity, in which the decision-making power of the basic industries, the drivers of order during industrialization, was gradually reduced. In reaction to the developing situation, the city councils and the political authorities, in a new alliance to overcome the consequences of the industrial decline, spread their influence over the Ruhr area, at first over the Hellweg zone, then over the Emscher zone. Different strategies of de-industrialization (economic shift vs. rationalization), an unequal availability of public subsidies and different planning approaches increased the overall pattern of organized complexity.

3 A Case of Socio-Economic Segregation and Organized Complexity: The City of Gelsenkirchen and the "Social City" Program

For more than a decade, various authors from Bochum university have developed ongoing research on the social areas in the Ruhr, thereby refining a set of criteria that is to give insights into various aspects of segregation and (social) deprivation (cf. Strohmeier and Kersting 1996; Strohmeier et al. 2001; Strohmeier 2003; Farwick et al. 2012). On the statistical level of the cities, Gelsenkirchen holds rank 1 in the 2000–2011 population losses, but contradictorily exhibits a relatively high proportion of under-18 population. At 14.1 % Gelsenkirchen (2012) also holds the highest rank of regional unemployment and consequently, next to the city of Duisburg, shows an extremely high proportion of low-income households; 21.7 % of the population live on social benefits which, again, is the highest score in the Ruhr area. Next to the city of Duisburg, an increasing 15.5 % share of foreign inhabitants is the highest in the region.

3.1 Organized Complexity Only Partly Revealed by Analytical Statistics

On the statistical level of the urban districts, the indicated overall multiple deprivation of Gelsenkirchen discloses some relative variations and relations indicating parts of the intrazonal complexity (Fig. 3). The demographic development is represented by the 2000–2010 population change (a). The rate of unemployment (b), the share of persons living on subsidies (c), and the population with residential property (d) refer to the economic situation. The ethnic segregation is reflected by the population share of foreigners and Germans with a migration background (e) as well as by the proportion of unemployed foreigners (f).

For the whole set of 18 urban districts and between all variables, correlation coefficients have been calculated. The social pattern presented thereby is mainly

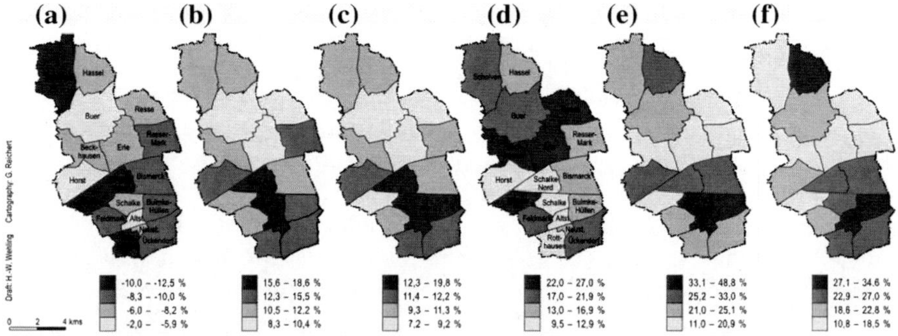

Fig. 3 Spatial patterns of social indicators in the city of Gelsenkirchen (*Data source* Statistikstelle Gelsenkirchen 2012)

determined by the economic variables (b) and (d). The rate of unemployment (b) strongly correlates with the proportion of persons living on subsidies ($r_s = 0.9804$) and with the proportion of foreigners and Germans with a migration background ($r_s = 0.7260$), which itself highly correlates with the persons living on public subsidies ($r_s = 0.7833$). High proportions of households with residential property define urban districts with low numbers of unemployed ($r_s = -0.8860$) and foreigners ($r_s = -0.8803$) as well as with low proportions of persons living on subsidies ($r_s = -0.8504$) and of unemployed foreigners ($r_s = -0.7276$). The ethnic poverty districts defined by the positive interrelations of the variables (b), (c), and (e) are Altstadt, Neustadt, Schalke, Schalke-Nord and Bulmke-Hüllen; at the opposite end of the range are the districts of Heßler, Erle, Resse and Beckhausen with high proportions of persons with residential property and the relative absence of persons that are unemployed, non-German and/or living on subsidies. Nine urban districts match with these two opposite constellations, the remaining nine ones can only slightly be connected to them. Variable (a) does not produce any direct high correlation with the other variables. As Farwick et al. (2012) show, it is the age structure that to a large extent triggers the population development. The age structure, however, can be affiliated to one of the defined groups; the younger population in the ethnic poverty districts cause a population decrease below average in these districts (Neustadt, Altstadt, Schalke, Hassel, Horst), whereas, disregarding the various variables of deprivation, the absence of younger households ends up in population losses above average (Resser Mark, Heßler, Scholven, Feldmark, Rotthausen, Ückendorf).

Therefore, according to the idea of organized complexity, the social variations shown in Fig. 3 could only in parts be recorded by means of basic analytical statistics. What could be shown was (1) that there are three types of districts, roughly addressed as rich, poor, mixed; (2) that there is a rich versus poor dichotomy indicating a self-organizing order that separates economic differences; and (3) that economically poor and mixed areas are separated mainly by non-economic (ethnic) differences. Hence, there are no more distinct zones, but rather a complex pattern of complex social systems.

3.2 *Can Urban Development Programs Change Patterns of Organized Complexity?*

If segregation goes along with discrimination and with steep gradients towards both the social and the spatial environment and if the self-organizing complex urban system seems to be anxious to marginalize both groups and places and if various marginalizing variables of segregation affect the same groups and places, critical conditions are likely. For example, with regard to health service and to education, the close interrelation of negative factors in the ethnic poverty districts is very likely to be perpetuated if the ethnic and economic segregation is not switched towards an appropriate participation in the urban society (Kersting et al. 2009; Farwick et al. 2012). Social deprivation will be prolonged into the future unless large parts of the next generation achieve a level of formal education that is sufficient to guarantee jobs on the labor market.

If self-organization in complex systems leads to a perpetuation of critical conditions, all hope rests on the effectiveness of urban renewal programs. To change and improve the presented situation and to enhance the perspectives, the integrated initiative 'Social City', later transferred into the program for experimental urban development 'ExWoSt—Stadtumbau West', was launched in 1999. Its aim is to stop the downward spiral in deprived quarters and comprehensively to improve the living conditions in order to stabilize the housing quarters demographically, economically, and socially by various means and objectives: The social and cultural infrastructure is to be adjusted to the local needs; the housing stock is to be renovated and modernized; house-owners with insufficient running means are supported financially to enable them to invest in their properties; the amount of free space within the fabric of land-uses is to be enlarged; the local economy is strengthened by developing qualification concepts for the insufficiently trained; the social infrastructure (schools, kindergartens, etc.) is adjusted in terms of quantity, quality, and sites; local resources and private investments are activated to manage the local districts and to improve their images. Various studies have shown that 'Social City', which pools the various modes of public funding and mobilizes the local resources and potentials, is an appropriate tool to combat social segregation and to stabilize deprived households and housing quarters (cf. Werner 2010).

The areas of Gelsenkirchen in which the initiative has been launched have been selected in a city-wide context and the social development objectives that are followed in them are in concordance with the overall urban development goals; it is not the aim of the initiative specifically to concentrate all activities on one area but to overcome the disconnection of this area from the overall urban development. The main criteria to select the areas and define their sizes are the level of unemployment, the dependency of the households on social benefits and the share of migrant households. In the city of Gelsenkirchen, this program has been or is being executed in the ethnic poverty districts of Schalke-Nord (until 2007), Altstadt, Neustadt, Bulmke-Hüllen (all since 2002) and Schalke (since 2008), in the shrinking district of Hassel (since 2009) and in the district of Bismarck (until 2007).

In accordance with principles of complex systems, this integrated approach regards urban renewal as a cross-sectional task to enable processes of change by creating networks and seeking synergies, e.g., among the city council departments. The program is monitored all over the city by continuous on-the-site discussion rounds that allow multifold forms of response. It has been learned that urban renewal of multiply deprived quarters in complex urban systems requires a multi-dimensional approach that has to be evaluated on the level of the urban district and on the background of the entire city; that city council departments may need to be re-organized; and that district offices that provide cross-sectional services for and keep incessant contact with the local population are the indispensable preconditions of a lasting result.

4 Conclusions

It could be shown that the southern and the northern parts of the Ruhr area have been affected by the urban-industrial development in different ways. One might argue that an industrial system spatially developed across the total Ruhr area. Exogenously directed by industrialists and economic politicians and trying to achieve an equilibrium of efficiency in terms of allocation, transport, and labor market, this system exhibited a disorganized complexity at first. In the Hellweg zone this system had to cope with an existing though still weak urban system of organized complexity, which until c. 1930 organized itself by adjusting to industrial needs, but since the interwar years has increasingly adjusted the industrial structures to the urban needs.

As there were no pre-industrial urban structures, patterns, or systems in the Emscher zone, the industrial system of disorganized complexity was able to unfold completely like a spatial circuit board printed on both sides of the river Emscher. Even if some (sub-)urban structures came into being, no self-organizing urban system able to cope with the spatio-industrial imprints was allowed to develop. It may have been particularly because of this that the Emscher zone was recognized as unfinished by contemporaries.

The collapse of the basic industries has to be regarded as a 'tipping point' of these complex systems, and the de-industrialization opened a new transition phase. This widely abolished the industrial system in the Hellweg zone and increasingly gave way to an urban system of organized complexity, initiated by the allied interests of the city councils and the national political authorities. In the Emscher zone, however, the industrial system, aligned in its disorganized complexity, prevailed for another two decades. Nevertheless, a nascent urban system of organized complexity gradually came into being. In this, the 1960s/1970s Emscher zone situation is similar to that of the 1930s Hellweg zone. From these decades on, one might begin to address the Ruhr as a whole as a complex urban system, but one has to recognize the still existing inequalities, the different maintenance capacities, and eventually the different qualities of the complexities of the two parts.

The de-industrialization not only had interregional phases in space and time, but it fundamentally changed the conditions for all systems from progress to decline which nearly immediately had socio-economic effects. At the end of various migration processes, multiple segregations appear in the Emscher zone, quite frequently located on the fringes of spatial patterns that have been part of the obsolete industrial system. The city of Gelsenkirchen has been taken to demonstrate the complexity formed by interrelated segregation variables.

To overcome the multiple deprivation, it seems to be necessary not to try to mitigate single aspects but, instead, to address the complex situation by an integrated development program. 'Social City' is regarded as an appropriate approach as it tries to apply system-related methods and to establish complexity-orientated institutions able to copy and correlate the organic evolution of urban systems. However, the 2012 Gelsenkirchen figures do not (yet) tell a bright story. This leads to further questions yet to be analyzed and answered: the question of the time spans of the development phases in urban complexity and the intensity/mechanisms of drivers necessary in order to change parts of the complex urban system in a sustainable way.

References

Batty, M.: Complexity in City Systems: Understanding, Evolution, and Design. Working Papers Series 117. Centre for Advanced Spatial Analysis (UCL), London (2007)

Batty, M.: The size, scale and shape of cities. Science **319**(5864), 769–771 (2008)

Farwick, A., Amonn, J., Groos, Th., Messer, A., Larsen, I., Teicke, M., Winkels, C.: Sozialraumanalyse Emscherregion, vol. 1. ZEFIR-Forschungsbericht, Bochum (2012)

Franke, E.: Das Ruhrgebiet und Ostpreußen: Geschichte, Umfang und Bedeutung der Ostpreußeneinwanderung, vol. 1. Volkstum im Ruhrgebiet, Essen (1936)

Gelsenkirchen: http://www.gelsenkirchen.de/de/Rathaus/Daten_und_Fakten/ (2012)

Jacobs, J.: The Death and Life of Great American Cities. Random House, New York (1961)

Kersting, V., Meyer, C., Strohmeier, K.P., Terpoorten, T.: Die A 40—der "Sozialäquator" des Ruhrgebiets In: Prossek, A. Schneider, H., Wessel, H.A., Wetterau, B., Wiktorin, D. (eds.) Atlas der Metropole Ruhr: Vielfalt und Wandel des Ruhrgebiets im Kartenbild, pp. 142–145. Emons, Cologne (2009)

Mehaffy, M.: The Kind of Problem Architecture Is: Jane Jacobs, Christopher Alexander and Since. New Science, New Urbanism, New Architecture? Towards a New 21st-Century Architecture. www.katarxis3.com/Jacobs.htm (2004)

Spethmann, H.: Das Ruhrgebiet im Wechselspiel von Land und Leuten, Wirtschaft, Technik und Politik. In: Ihde, G., Wehling, H.W. (eds. and comm.) Klartext, vols. 4 and 5. Essen (2011 [orig.: 1935–1940])

Steinberg, H.G.: Das Ruhrgebiet im 19. und 20. Jahrhundert: Ein Verdichtungsraum im Wandel. Siedlung und Landschaft in Westfalen. 16. Selbstverlag der Geographischen Kommission für Westfalen, Münster (1985)

Strohmeier, K.P.: Sozialraumanalyse. In: Institut für Landes- und Stadtentwicklungsforschung des Landes Nordrhein-Westfalen, ILS (ed.) Raumbeobachtungssysteme als Instrumente der integrierten Stadt(teil-)entwicklung, pp. 22–33. Self-published, Dortmund (2003)

Strohmeier, K.P., Kersting, V.: Sozialraum Ruhrgebiet—stadträumliche Differenzierungen von Lebenslagen, Armut und informelle Solidarpotentiale. In: Bovermann, R., Goch, S., Priamus,

H.J. (eds.) Das Ruhrgebiet—Ein starkes Stück Nordrhein-Westfalen, pp. 451–475. Klartext, Essen (1996)

Strohmeier, K.P., Prey, G., Heidbrink, I.: Sozialraumanalyse Gelsenkirchen, stadträumliche Differenzierungen von Lebenslagen und Lebensformen der Bevölkerung, Armut und politischer Partizipation. Materialien und Analysen zur Begründung der Auswahl eines Stadtteils mit besonderem Erneuerungsbedarf. Abschlussbericht. ZEFIR—Zentrum für interdisziplinäre Ruhrgebietsforschung, Bochum (2001)

Weaver, W.: Science and complexity. Am. Sci. **36**(4), 536–544 (1948)

Werner, S.: Evaluating the Complexity of the "Socially Integrative City". www.difu.de/node/6994 (2010)

An Organizational Cybernetics Approach to University Planning in an Urban Context: Four Intervention Experiences

Xosé L. Martínez Suárez and José Pérez Ríos

Abstract In this chapter we discuss how the use of Organizational Cybernetics (OC) helps to identify the variety of needs to be considered in a planning-related intervention in a university campus and to influence the project design in such a way that the variety of services responding to these needs will also be increased. After a short description of some OC concepts and of the systemic methodological framework that has been used in this work, we present examples of application of this framework in four urbanistic and architectonic interventions on the UDC campuses.

Keywords Organizational cybernetics · University urban planning · Viable system model · VSMod · Recursion levels-key factors matrix · University of a Coruña

> It is taken for granted that in actual design form should be used to reinforce meaning, and not to negate it (K. Lynch 1960, p. 46).

1 Introduction

The aspects that may have an influence in urban planning interventions can be highly diverse (i.e., related to demographic, legislative, economic, ecological, sociological, architectonic, urbanistic, academic, transport domains, etc.). In order

X. L. Martínez Suárez
University of A Coruña, E.T.S. Arquitectura, Campus A Zapateira, 15071 A Coruña, Spain
e-mail: xlms28@yahoo.es

J. P. Ríos (✉)
University of Valladolid, E.T.S.I. Informatica, Campus Miguel Delibes, 47011 Valladolid, Spain
e-mail: rios@uva.es

C. Walloth et al. (eds.), *Understanding Complex Urban Systems: Multidisciplinary Approaches to Modeling*, Understanding Complex Systems,
DOI: 10.1007/978-3-319-02996-2_7, © Springer International Publishing Switzerland 2014

to cope with the complexity generated by the potential interrelations among these domains, appropriate approaches and methodological tools are needed. In this work we present an example of how a public university, the University of A Coruña (UDC) in Galicia, Spain, faced this complexity by means of a systemic methodology.

Since we are using the term ‘complexity’ to indicate the level of difficulty in dealing with a certain situation, it may be convenient to clarify its meaning in our context. Among the various concepts of complexity, one finds computational, technical, organizational, personal and emotional complexity (Yolles 1999), static, structural and dynamic complexity (Cambel 1992), the latter of which can clearly be distinguished from detail complexity (Senge 1990). Here, in the context of urban planning in a university environment, we will use the term to characterize the potential of a system for taking multiple shapes, where shapes are not limited to spatial or physical structures, but may refer to any state of any aspect of the system under consideration today or in the future. In the case of social systems, that potential also includes possible forms of behavior. Hence, the complexity of the system can be expressed in terms of the multiplicity of states it can take, which is variety, or, in the words of Beer (1979): “The measure of complexity is called variety and variety is defined as the number of possible states of whatever it is whose complexity we want to measure” (p. 32). This is the definition of ‘variety’ used in the context of this paper.

Once the meaning of variety as a measure of complexity is clarified, another aspect to take into account is how to deal with complexity. At this point it may be convenient to recall the Conant and Ashby Theorem (1970), which states: “Every good regulator of a system must be a model of that system”, and its implication as Schwaninger (2010) puts it: “[T]he results of any management process cannot be better than the underlying model except by chance” (p. 1421). If we consider an urban planning intervention in university contexts as a complex task, we should use adequate models to carry it out. Those models should help university decision makers to deal with the complexity (variety) faced. One such model is Beer’s Organizational Cybernetics (OC) approach, which was created, among other reasons, precisely to help decision makers deal with variety by, e.g., diagnosing the viability of an existing system.

The Conant and Ashby Theorem further implies that those models should have the capacity to provide decision makers with the variety needed to deal with the variety they face. This is, what Ashby’s Law of Requisite Variety (1956) means: “only variety can destroy variety” (p. 207) (in the sense of ‘absorbing’ variety, according to Beer). Its implication is that when we try to design a new system that is intended to meet certain aims in a dynamic environment, we first have to identify the variety that the system must face. In the case of an architectonic and urban planning intervention, this would mean identifying the variety of needs to be satisfied by the intervention. Once these needs are recognized, it is necessary to design the system that is going to satisfy them. This means that the system has to be able to generate an amount of variety that is in consonance with the identified needed variety.

However, a system with enough capacity to satisfy the variety of needs is not sufficient if it does not make efficient use of this variety. This means that an organization has to be created (or an existing one adapted) that will be in charge of governing (managing) the functioning of the system. A final requisite will be that the system designed must continue to satisfy needs as they change over time. Hence, the system designed should have the capacity to adapt and, ideally, evolve in response to the changes happening in the environment, which yield different or new needs. In summary, the system should be designed in such a way that it is able to meet its purposes not only at present but also in the future. The ability of Beer's OC model to help design such a system (e.g., an organization) was among the reasons for its application in the university urban planning context.

To deal with the complexity faced by an organization (in our case the UDC) that tries to reach a goal in a certain environment (the geographical area in which the UDC has or may have an impact), OC has various tools to offer. One of them helps reduce the variety of the problem to be tackled (simplification) while another one helps match or increase the variety deployed by the organization. Since the context of designing a system (or, in our case, an urbanistic or architectonic intervention to redesign a system) can be interpreted in the sense that it is necessary first to identify the variety of the situation to be dealt with and, next, to deploy a similar or larger amount of variety through the interventions conceived, we particularly used the second option, trying to cover as many identified needs as possible.

In order to increase variety we related the university architecture to the surrounding urban space, streets and squares. Hence, the approach chosen goes far beyond a spatial container model of the university campus. We also tried to overcome the simplistic planning approach applied in many university campuses by analyzing the existing variety. Hence, the OC and urban planning practice were intertwined in the urban project with the aim of transcending a closed and fragmented conception of the university space.

Summarizing these considerations, the purpose of this paper is to show how combining the OC approach with university urban planning knowledge can be used to increase variety. We hypothesize that the use of OC helps to identify the variety of needs to be considered in a planning-related intervention in a university campus and to influence the project design in such a way that the variety of services responding to these needs will also be increased.

The paper is structured as follows: We start with a brief introduction to the University of A Coruña, the setting of the case studies. This is followed by a short description of some of Beer's OC concepts—which we will only introduce to the extent relevant for an urban planning intervention in university contexts—and of the systemic methodological framework that has been used in this work. The next part of the paper is dedicated to examples of application of this framework in four urbanistic and architectonic interventions on the UDC campuses. The paper ends with some conclusions.

2 The University of A Coruña's Renewed Strategy

The UDC is one of the three public universities that compose the university system of the Autonomous Community of Galicia (Spain), the other two being Santiago de Compostela and Vigo. The UDC, with its approximately 24,000 students, has branches in the cities of A Coruña (five campuses) and Ferrol (two campuses), as shown in Fig. 1. A distance of 60 km separates both cities, forming an urban region populated by around 650,000 inhabitants.

The two campuses in the city of Ferrol, Esteiro and Serantes, consist of 12 and 1 buildings, respectively. The campuses in the city of A Coruña, Elviña-A Zapateira and Riazor (Fig. 2), are composed of 19 and 2 buildings respectively.

The long-term strategy project, some architectonic and urbanistic interventions which this paper refers to, started in 2004, when a new rectorate redefined the UDC's mission and purposes. The rectorate went so far as to challenge the fit of the campuses' architectonic and urbanistic aspects with the redefined UDC's purposes. A guiding question was how universities, through their governing policy actions, might overcome the divide between university and city implicit in many campuses from the second half of the twentieth century.

The initial hypothesis of the vice-rectorship of Infrastructures and Sustainable Development was that variety in place did not match the requisite variety implied by the UDC's redefined purposes. This hypothesis led to the vice-rectorship's decision to devise strategies and actions to increase the variety related to architectonic and urbanistic aspects (Pérez Ríos and Martínez Suárez 2007, 2011;

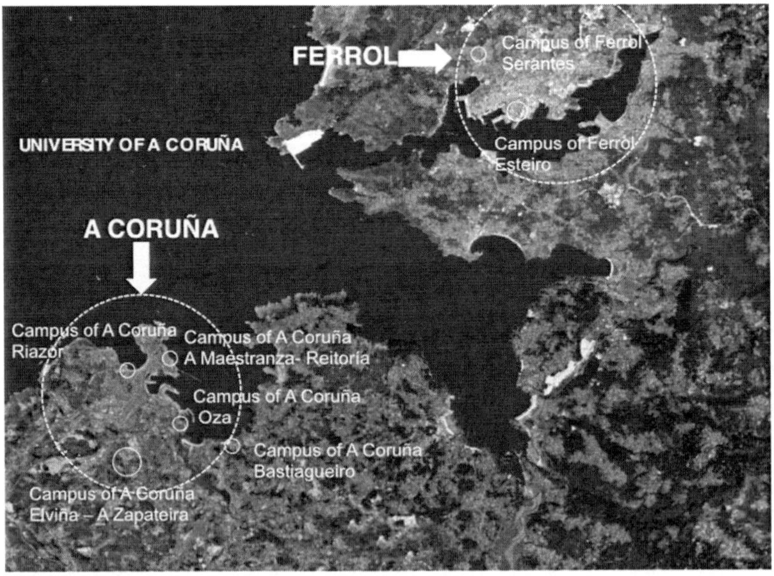

Fig. 1 UDC campuses in the cities of A Coruña and Ferrol

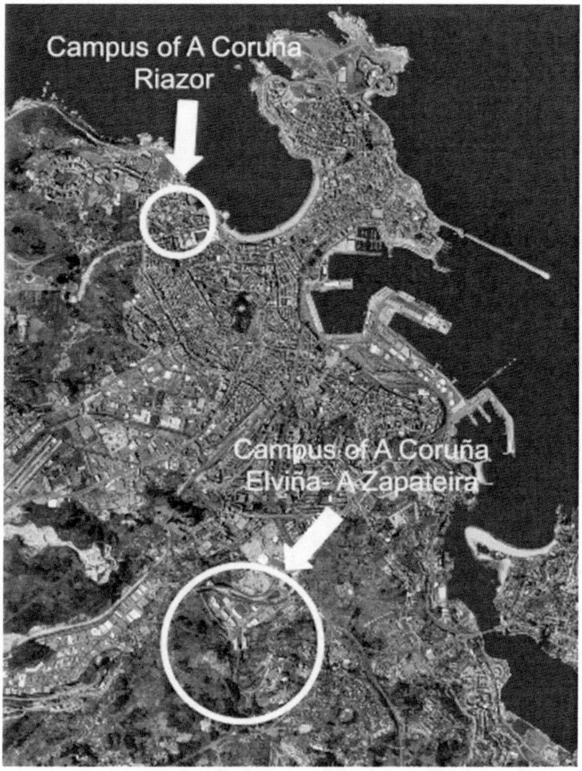

Fig. 2 UDC campus of Elviña-A Zapateira and campus of Riazor (city of A Coruña)

Martínez Suárez 2008, 2010) aiming at, for example, an increase of services offered by means of reconfigured university spaces.

To choose and detail these interventions, the OC approach was selected because of its capacity to deal with variety. Several interventions were carried out and, e.g., the relation between the buildings (faculties, service centers), their environments and the city quarters in which they are embedded or to which they were adjacent was addressed.

3 An Organizational Cybernetics Approach

3.1 Organizational Cybernetics and the Viable System Model

As we have already mentioned, the approach applied in the case of the UDC is related to Beer's OC. This approach was chosen for its capacity to design and/or diagnose organizations (systems) and to deal with variety (complexity) (Espejo and Harnden 1989; Espejo and Reyes 2011; Schwaninger 2006; Schwaninger and Pérez Ríos 2008; Pérez Ríos 2008, 2012). The main elements of OC and the details of the systemic methodological framework used in the UDC case are described elsewhere (Pérez Ríos 2010, 2012). Here, we will introduce the approach only in so far as it is necessary to understand the UDC's interventions.

In addition to the concepts of variety, Ashby's Law, and the Conant-Ashby Theorem already mentioned above, we consider the OC concept of viability relevant to our case. Viability refers to the capacity of a system to maintain a separate existence, that is, to survive regardless of changes in its environment. For that, it must have the capacities of self-regulation, learning, adaptation, and evolution.

In his Viable System Model (VSM), Beer (1981, 1985) establishes the necessary and sufficient conditions for the viability of an organization. These are related to the existence of a set of functional systems (Beer identified them as System 1, 2, 3/3*, 4, and 5) in an organization and a set of relationships among these functional systems and the environment. According to Beer, all viable systems contain viable systems and are themselves contained in viable systems. The most important aspect of this recursive conception of viable systems is that, no matter which place they occupy within the chain of systems, they must always, in order to be viable, contain the five functional systems that determine viability.

Based on OC and, in particular, the VSM's conceptual elements, Pérez Ríos (2010) introduced a systemic methodological framework to help design or diagnose systems in view of their viability. This is the framework used in the UDC case. The following will provide an outlook on its main components and on how it works.

3.2 A Systemic Methodological Framework

The process for designing or diagnosing an organization that is supposed to attain a certain purpose can be structured into up to four stages, depending on the level of insight required before interventions can be carried out. In the case of the UDC, the first and second of these four stages were applied. The first stage is related to the identification and clarification of the organization's identity and purpose. The answer to questions such as what the organization is (and what it is not) and what its purposes are, also help to delimit the boundaries of the organization. In a second stage the vertical structure of the organization is created or identified. Creating this vertical structure by vertical unfolding according to the concept of "complexity unfolding" (Espejo 1989) helps to deal with the complexity faced.

The process of vertical unfolding implies dividing the system into sub-systems and these into sub-sub-systems and so on, just as in Beer's recursive conception of the VSM. The result of this process applied to an organization will be a set of sub-organizations, sub-sub-organizations, etc., each of which will have to deal only with their respective environments (Fig. 3).

It is important to keep in mind that we do not reduce complexity through this vertical unfolding process. Vertical unfolding differs from the reductionist-analytical approach, by which the system under study is divided into parts to be studied

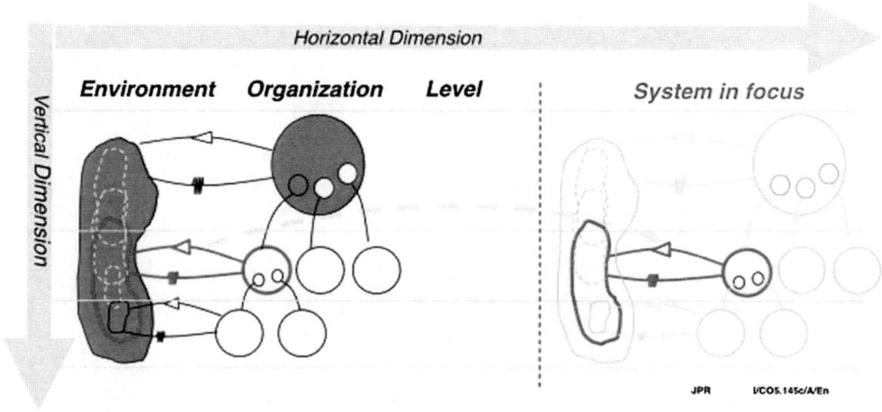

Fig. 3 Vertical and horizontal dimensions of the system under study (Pérez Ríos 2008)

separately. By vertical unfolding, we 'zoom in' on the sub-organizations and their environments.

Those designing or diagnosing the system under study determine the recursion criteria as well as the number of recursion levels, i.e. the criteria along which and the depth to which vertical unfolding is carried out. The use of various recursion criteria allows for a more comprehensive study of an organization.

Once the vertical levels are defined, the purposes and key factors of each level are identified. Since each level is embedded in the previous one, the purposes, at all levels, must be recursively coherent up to the general purpose of the whole organization at the initial level. The number and kind of key factors to be taken into account is defined in each specific situation by those responsible for the study. Only two or three key factors might be considered necessary, or many more. Key factors may also be different for different recursion levels, which again allows for a more comprehensive study of the organization in question. As we will see in the next section, in the case of the UDC we have determined six key factors.

A tool to visualize the resulting structure is the Recursion Levels-Key Factors Matrix (Pérez Ríos 2008). The vertical axis of the matrix shows the recursion levels and the horizontal axis shows the key factors. Each matrix represents the vertical unfolding along one recursion criterion (Fig. 4). Since the visualization of the several matrices and the switching between them can hardly be carried out on paper, the specialized software VSMod®, created specifically to this end, can be applied (Pérez Ríos 2003, 2012).

As we indicated in the introduction to this article, our hypothesis was that the use of an OC approach helps identify the variety of needs to be considered in a planning-related intervention in a university context and to influence the project design in a way so that the variety of services responding to those needs will also be increased. Based on the results of the two-stage process of the systemic methodological framework, several architectonic and urbanistic interventions were

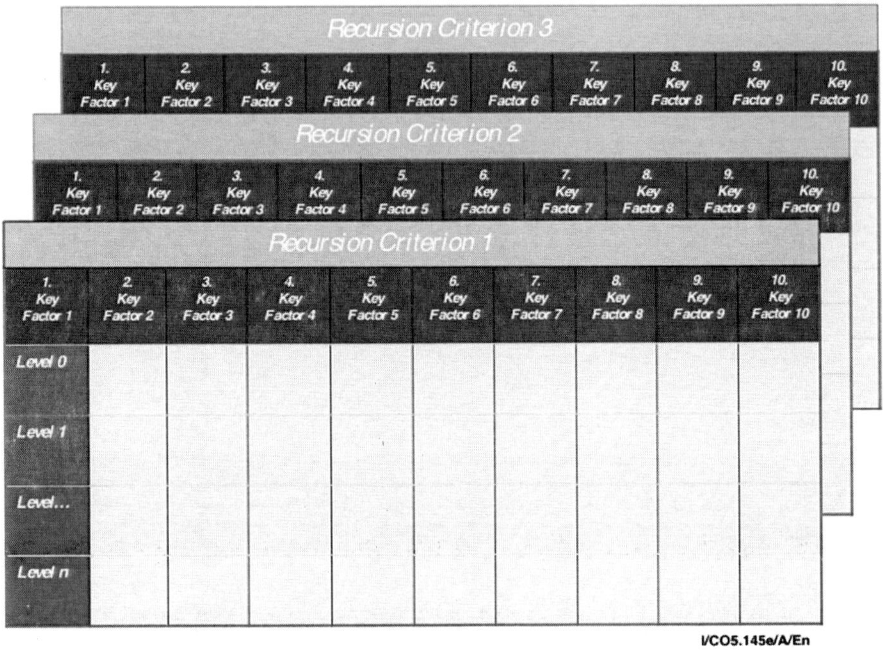

Fig. 4 Recursion Levels-Key Factors Matrices. The *rows* show recursion levels, the *columns* show key factors, and each matrix corresponds to a recursion criterion (Pérez Ríos 2008)

identified, four of which—chosen from the same recursion level to assure comparability—are described below.

4 The Case of the UDC

4.1 Spatial Unfolding of the UCD's Purpose

According to the systemic methodological framework, in a first stage the UDC's identity, mission, purposes, and also its spatial area of influence were clarified (Pérez Ríos and Martínez Suárez 2007). In a second stage, the vertical unfolding was carried out.

As outlined above, some of the UDC's redefined purposes implied a closer look at architectonic and urbanistic aspects of the campuses. This, in turn, required a definition of the geographical region affected by the UDC campuses, which is the urban region of A Coruña-Ferrol described above.

In line with the results from the first stage, in the second stage we established the spatial (geographic) dimension of the university as the first recursion criterion of the vertical complexity unfolding. Five recursion levels, in this case spatial

1. Recursion Level	2. Spacial Scope	3. Relevant Issues/ Purpose	6. Influencial Institutions/Organisms	7. Applicable Legislation	8. Actions Formulated
0	Galicia.	- Social Function of the universities. - Relationship with the urban policy. - University Housing Policy.	- Xunta de Galicia. -Ministries: Education, Territorial Policy, Housing, Environment and sustained development. - Universities: A Coruña.; Santiago de Compostela; Vigo.	1. Act 10/1995 on Town and Land Planning of Galicia. 2. Ground/Building Act of Galicia (December 2002). 3. Act 11/1989 on Galicia University System Planning. 4. University Act 6/2001 5. UDC standing Rule	- Contribution of the UDC to the Town-Planning Guidelines in Galicia (in progress) - URB 9. Scientific-Technological Park. - URB 16. University Residential Area (Campus Elviña). - URB 15. Research Area. Creation of new enterprises.
1	Urban Region A Coruña Ferrol	- Accessibility. - Range (number of potential students) - Visibility of the UDC in the cities, small towns and villages. - Economic and social development of the urban region. - Connection with the business network.	- RENFE.(Spain's Railway System) - Cities: A Coruña; Ferrol and all the rest in the Urban Region. - UDC. - Xunta de Galicia. (Commuting)		- URB 1. Territorial Accessibility: shire Public Transport Suburban Trains, and coach network. - URB 12. Parking Lots - Parking Lots at Railway Stations. - URB 13. Bus, Train Station Campus Elviña. - URB 11. Campus Center - URB 18. Intermodal Station
2	a) Urban A Coruña. b) Urban Ferrol.	- Accessibility. - Integration University/city - Cohesion university/city. - Structuring of public equipment and urban services with the university.	- UDC. - City of A Coruña. - City of Ferrol.	- Urban Master plan of A Coruña (1995). - Urban Master plan of Ferrol.	- URB 2. Enlargement of Urban Coaches network. - URB 17. Bicycle lane pedestrian Path from the city-centre to the campus.
3	a) Campus A Coruña	- Adaptation to the European Union directives on Universities degrees. - Urban attraction - Urban and architectonic referent (model of sustained development). - Functional complexity.	- UDC - City of A Coruña.	- Urban Plan for Elviña-A Zapateira Campus (1991) and its modification in 2002. - Environment Plan.	- URB 10. Area 30. Elviña Campus coach. - URB 9 Scientific-Technological Park. Botanical Park. (Campus focus) - URB 19. Ecologic Quarter.
4	Single Buildings.	- Functionality. - Comfort and Environment Managing. - Optimizing spaces. - Reference on sustainability.	- UDC - Institution Board. - City of A Coruña. - City of Ferrol.	- Urban Master plan of A Coruña (1995). - Special Plan Campus of Esteiro.	Actions at each particular centre. - URB 8. Redesign of Zapateira Square. - URB 20. Riazor Campus - URB 21. Ferrol-Esteiro Campus

Fig. 5 Recursion Levels-Key Factors Matrix (v.1.1) for UDC

levels, were identified: the autonomous community of Galicia, the urban region of A Coruña-Ferrol, the urban areas of A Coruña and Ferrol, the campuses of the UDC, and the individual (faculty or service) buildings and spaces on each campus. Each spatial level is represented by a row in the Recursion Levels-Key Factors Matrix (Fig. 5).

For each recursion level we identified the sub-system's specific purpose, i.e. we applied the first stage of the systemic methodological approach to each sub-system while performing the vertical unfolding according to the second stage. This identification of a purpose (as well as an identity and boundary) for each sub-system reflects the viability criterion for (sub-)systems. All purposes of all sub-systems at all levels are recursively coherent up to the general purpose of the UDC. Furthermore, for each recursion level we identified its specific sub-environment and the key factors that related to, e.g., town-planning legislation and administrative structures. Finally, based on the purposes and key factors identified on each recursion level, we suggested architectonic and urbanistic interventions to be carried out (denoted URB 1 to 21 in Fig. 5).

If the systemic methodological approach was indeed helpful in devising interventions that support the identified purpose of the system, then, in the UDC's case, the interventions would need to increase the variety of uses of the university and adjacent urban space. In order to support our initial hypothesis on the usefulness of an OC approach, we will present four distinct interventions based on the systemic methodological approach. We will also review each intervention's impact on the increase of variety on the level of individual buildings and spaces.

4.2 Case 1: A Fraga

The first intervention that we provide as an example concerns the A Fraga area of the Elviña-A Zapateira campus, which is located at the periphery of A Coruña. The area was built between 1970 and 1990, with buildings organized around a central open space.

The central open space, due to its proximity to all entrance doors, had been the preferred parking area of students and staff members alike since 1974. However, conflict and tension had grown between car owners and pedestrians, leading to an aggressive traffic situation.

In line with the general aim to increase the variety related to architectonic and urbanistic aspects of the UDC campuses, the target for the A Fraga area was to recover the central open space as a centre for university citizenship. A working hypothesis to reach this target was that the relationships between academic and social functions needed to be increased.

In order to thoroughly devise the intervention, a closer look at the identity, purpose and boundaries of the recursion level of individual (faculty or service) buildings and spaces was required. Hence, a workshop was organized in 2004 with the goal to identify the identity, purpose and boundaries of the central open space and its environment. The three-day-workshop was open to the university collective (i.e. students and staff members) studying and working in the Architecture Technical Schools and Biology and Chemistry Faculties surrounding the central open space. It had been called in under the general topic of "University public space and sustainability" and attracted more than 100 students and staff members.

The discussion that finally led to a better understanding of the identity, purpose and boundaries of the central open space was stimulated by rather general questions. One such question concerned the discrepancies between the good practices taught in class and the actual use of the central open space. Another question addressed how the university could make use of academic knowledge to become a transformation agent and a cultural reference through architectonic and urbanistic practices in the city and in the territory. The overarching theme was, of course, the absence of relationship spaces between the buildings surrounding the central open space.

Through this participative workshop, the different purposes of the central open space perceived by different users could be accounted for and included in the plans to increase the variety of this space. Thus, we include the views of observers in different roles into the study, an issue discussed in what is known as second-order cybernetics by, e.g., von Foerster (1995).

Following this workshop, considering the individual purposes perceived for the central open space as well as the overarching purpose of the UDC in general, recommendations for actions were formulated and provided for the decision-making process. As a consequence, the central open space was transformed into a large civic area in order to reflect the different functions of the various buildings and relate them to each other. This also required a partial redesign of buildings,

Fig. 6 Action URB 8: redesign of the central open space at A Fraga (A Zapateira-Elviña): Before (*left*) and after (*center* and *right*) the intervention

connecting their ground floors with the outside space, thus promoting possible new private (commercial) and public uses. The location to park private cars was moved to a larger parking area 200 m away.

The images in Fig. 6 demonstrate the strong visual effect of the spatial transformation. Indeed, the central open space now relates the surrounding buildings to each other. The space became a meeting point for students, a green area for relaxation, a venue for university events, a space for occasional lectures, as well as a recreation area with exceptional views over the city of A Coruña.

Hence, the functional complexity, i.e. the variety of uses of the A Fraga area, increased through an intervention informed by the use of the systemic methodological framework. The systemic methodological framework had been applied to identify the A Fraga sub-system of the UDC and its specific purposes. In particular, the participative component helped to identify the different purposes of the central open space perceived by different observers. An intervention was devised, carried out, and succeeded in increasing the variety of the place, which is in line with the overall purpose of the UDC.

4.3 Case 2: Riazor

The second example concerns the campus of Riazor in the central urban area of A Coruña. A. Tenreiro designed this campus in the years just prior to the Spanish civil war in 1936. The initial drawings showed an exterior space understood as an open, educational and cultural space. The adaptation of the plan started in the 1950s. It produced buildings and spaces separated from the immediate urban environment by high walls and fences (Fig. 7). Paved parking areas characterized the enclosed campus spaces. Hence, Riazor provided a typical example of a divide between university and city.

The potential purpose of the Riazor campus could directly be derived from the redefined purpose of the UDC. It was immediately clear that relations between the buildings and spaces of Riazor campus and the adjacent streets and city quarters had to be established. Hence, any intervention was to be aimed at making

Fig. 7 Riazor. Removal of architectonic barriers (walls). Before the intervention

Fig. 8 Riazor. Removal of architectonic barriers (walls). Old School of Commerce (*right*)

accessible the university space to the city, at opening up show rooms and study rooms, as well as external spaces for public use. Identified key factors related to the civic use of space and the historical meaning of buildings—the latter both in the sense of buildings seen as a materialization of a citizen's image of the city and in their role of recuperating the memory of long-planned physical development and, hence, in the role of connecting the urban past and future.

On April 14, 2011 the UDC removed a wall (Fig. 8), 100 m long and 3 m high, uncovering a fine example of earlier architecture, the old School of Commerce. Also, the buildings' interiors were adapted to new uses and made accessible to the public: show rooms, study rooms, and a seniors' university. The restricted parking area was converted into a green space open towards the street and adjacent city quarters.

The intervention at the Riazor campus significantly increased the variety of the university-city relationship in both expected and unexpected ways. Expected and aimed at was the establishment of an historically meaningful urban space and the opening up of the campus to the public. In particular, with the removal of the wall the relation between the old School of Commerce and the other university buildings was re-established and an urban landscape was recovered for the city that had already been conceived in the original project from 1936, facilitating interactions and a variety of new uses. Unexpectedly, during the process of wall demolition, new urban relations where uncovered. In the short term, a dialogue between buildings became possible and, in the long run, the most representative city building—the Hercules Tower, a World Heritage Site—became a reference point for the university space and city streets (Fig. 9).

Fig. 9 Riazor. Architectonic barriers removed and city symbol

Thus, the variety of the urban environment has been significantly increased by the intervention. Led by the UDC's and its subsystem's purposes identified by the application of the systemic methodological framework, a closed urban space with virtually no variety of use was transformed into a new green area, a space common to university members and citizens, a cultural space, a space pointing out historical symbols—the original university project and the Hercules tower—, a space of new services offered to citizens, etc.

4.4 Case 3: Esteiro

The third example of an intervention at the UDC informed by the application of the systemic methodological framework refers to the establishing of the Esteiro campus in the city of Ferrol. This campus has been created by the transformation of a centrally located former military hospital. Though thoroughly refurbished, the building kept its architectonic characteristics.

Despite the transformation, the former military area remains closed to the city. Access to the campus gardens is restricted by means of a metal fence recently built, creating an initial situation similar to the one of the previous example (Riazor campus in A Coruña). Obviously, in the light of the UDC's purposes and the aim to increase variety by means of, e.g., increased urban connectivity, fencing a university campus poses a problem.

Fig. 10 Esteiro, Ferrol. Before (*left*) and after (*center* and *right*) the intervention

The systemic methodological framework was applied with similar results as for the case of Riazor. The purpose of the university space was identified as congruent with the UDC's purpose to overcome the divide between university and city. Key factors identified in the second stage were related to the public perception of the campus and to attractive architectonic-urban form that inspires people to enter it.

Consequently, a proposal was submitted to the Ferrol municipality and the faculties' management to modify the urban planning of the campus. Where there were plans for a new building, it was proposed to remove the metal fence and to create a new public space, a square that could serve as a relation space between the campus and the neighborhood.

The suggestions were implemented in a new construction dedicated to various services (cafeteria, cultural activities rooms, gymnasium, exhibition spaces, library, etc.) not only for students, but also for the general public (Fig. 10). It is further intended to establish a relational space by expanding the square in front of the new building beyond the university plot limits and to dedicate its use to pedestrians.

Hence, this intervention, similarly based on the application of the systemic methodological framework, has already led to increasing the variety of uses by opening up the campus of Esteiro and providing services to the neighborhood. Further interventions can be based on the suggested actions.

4.5 Elviña

The fourth and last example we provide is of an intervention at the Elviña campus in A Coruña. This was triggered by the Faculty of Law's request to adapt its building and facilities to the European Higher Education Area requirements. This adaption was taken as an opportunity to transform both the interior and the exterior of the building. The 1991 original design concept for this campus was based on a systematic repetition of a rectangular module-type building on land plots approximately 130 m long and 60 m wide. These buildings of three or four floors were completely enclosed by roads and parking areas. The urbanistic 'insularity' of each plot was aggravated by a steep topography and the architectonic form selected for the design o the buildings (Fig. 11, left).

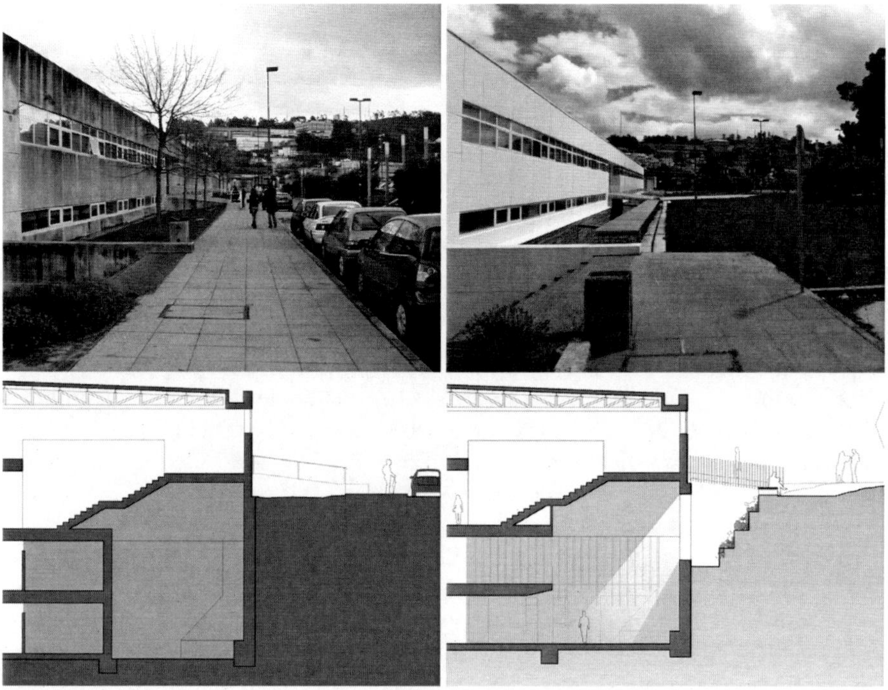

Fig. 11 Elviña campus-Faculty of Law. Before (*left*) and after (*right*) the intervention. (Images AE Estudio)

The violation of the UDC's aims to increase variety by means of, e.g., establishing relations in the architectonic and urbanistic environment, is obvious in this example. Buildings were detached from each other as well as from connecting walkways, impeding most interaction among members of different faculties. The intervention suggested—based on the application of the systemic methodological framework and making use of the window of opportunity when adaptation was requested—to break up the buildings' insularity and to establish relations between the main building's façade and the surrounding environment. This suggestion has been implemented by creating a public space in front of the Faculty of Law's building shifting the road from 5 m to more than 20 m away from the façade and by redesigning the building's interior (Fig. 11). These were the objectives given to the architects (AE Estudio, A. Martín Prieto–E. Sarmiento, architects). This action will be replicated in the Faculty of Law's two adjacent buildings, so these three buildings will be connected by means of newly created common public space.

This intervention, devised based on the application of the systemic methodological framework and the spatial unfolding of the UDC's purpose, aimed at and successfully managed to transform what was a set of isolated buildings with no communication among them and with poor relation opportunities among university members (low variety) into an interrelationship space. Paths of communication—both visual and material—have been created, thus increasing the variety of uses of the campus space.

5 Conclusions

All four interventions described above were guided by the use of the systemic methodological framework to identify actions suited to the aim of increasing the variety of university campuses by means of architecture and urbanism. As part of the systemic methodological framework, the Recursion Levels-Key Factors Matrix helped to identify the specific purposes and key factors leading to the deployment of the variety required, i.e. Ashby's requisite variety. In particular, the participative workshop applied in the A Fraga intervention helped to identify the various purposes of an open space perceived by different observers.

What is common to the four interventions described above is that they actually succeeded in increasing the complexity (variety, in cybernetic terms) of relations on the respective university campus and with the surrounding urban environment.

We can therefore conclude that applying the systemic methodological framework—an OC approach—helped us to identify the variety of needs for each intervention, to consider these needs in the redesign of physical spaces and in the transformation of buildings, and to increase the variety of services provided.[1]

References

Ashby, W.R.: An Introduction to Cybernetics. Chapman & Hall, London (1956)
Beer, S.: The Heart of Enterprise. Wiley, Chichester (1979)
Beer, S.: Brain of the Firm, 2nd edn. Wiley, Chichester (1981)
Beer, S.: Diagnosing the System for Organizations. Wiley, Chichester (1985)
Cambel, A.B.: Applied Chaos Theory: A Paradigm for Complexity. Academic, New York (1992)
Conant, R.C., Ashby, W.R.: Every Good Regulator of a System Must Be a Model of that System. Int. J. Syst. Sci. **1**(2), 89–97 (1970)
Espejo, R.: The VSM revisited. In: Espejo, R., Harnden, R. (eds.) The Viable System Model: Interpretations and Applications of Stafford Beer's VSM, pp. 77–100. Wiley, Chichester (1989)
Espejo, R., Harnden, R. (eds.): The Viable System Model: Interpretations and Applications of Stafford Beer's VSM. Wiley, Chichester (1989)
Espejo, R., Reyes, A.: Organizational Systems: Managing Complexity with the Viable System Model. Springer, Heidelberg, Berlin (2011)
Lynch, K.: The Image of the City. The MIT Press, Cambridge (1960)

[1] This work has been supported in part by the Ministerio de Ciencia e Innovación (Plan Nacional de I+D+i) of Spain. Ref.: CSO2010-15745. The authors also wish to thank Christian Walloth for his thoughtful and precise suggestions.

Martínez Suárez, X.L. (Dir.): Areas Supramunicipais da Provincia da Coruña. Investigación urbanística. Diputacion Provincial da Coruña (2008)
Martínez Suárez, X.L.: Estratexia de aloxamento universitario e UDC. In: D. Hooper et al. (eds.) Areas residenciais universitarias. Concello da Coruña (2010)
Pérez Ríos, J.: VSMod®: A Software Tool for the Application of the Viable System Model. 47th Annual Conference of the International Society for the Systems Sciences (ISSS), Heraklion (2003)
Pérez Ríos, J.: Diseño y diagnóstico de organizaciones viables. Un enfoque sistémico. Iberfora 2000 (2008)
Pérez Ríos, J.: Models of organizational cybernetics for diagnosis and design. Kybernetes **39**(9/10), 1529–1550 (2010)
Pérez Ríos, J.: Design and Diagnosis for Sustainable Organizations: The Viable System Method. Springer, Berlin, Heidelberg (2012)
Pérez Ríos, J., Martínez Suárez, X.L.: Applying VSM in the strategic management of A Coruña University in Galicia, Spain. In: Christopher, W.F. (ed.) Holistic Management: Managing What Matters for Company Success, pp. 124–134. Wiley, Hoboken, New Jersey (2007)
Pérez Ríos, J., Martínez Suárez, X.L.: Professional systemics in urban planning: the case of the University of A Coruña. Keynote speech. 7th National and International HSSS Conference, Athens (2011)
Schwaninger, M.: Intelligent Organizations: Powerful Models for Systemic Management. Springer, Berlin, Heidelberg (2006)
Schwaninger, M.: Model-based management (MBM): a vital prerequisite for organizational viability. Kybernetes **39**(9/10), 1419–1428 (2010)
Schwaninger, M., Pérez Ríos, J.: Organizational cybernetics in focus. Int. J. Appl. Syst. Stud. **2**(1/2), 1–5 (2008)
Senge, P.M.: The Fifth Discipline: The Art and Practice of the Learning Organization. Doubleday/Currency, New York (1990)
Von Foerster, H.: Cybernetics of Cybernetics, 2nd edn. Future Systems Inc, Minneapolis (1995)
Yolles, M.: Management Systems: A Viable Approach. Financial Times Pitman, London (1999)

Emergence in Complex Urban Systems: Blessing or Curse of Planning Efforts?

Christian Walloth

Abstract Today's urban planning practice largely relies on the assumption that the same interventions create the same results in comparable cities. However, in complex urban systems, where emergent, i.e. unplanned and in principle unpredictable, qualities lead to successive qualitative changes of the environment, the repetition of interventions does not guarantee the same results. Where emergent qualities cannot be foreseen or planned, and where the outcome of urban interventions depends on emergent qualities, urban planning practice has to widen its scope towards emergent qualities. In this vein, the present article discusses the potentially significant impacts of emergent qualities using examples from non-capital cities in Central and Eastern Europe. On the one hand, this article calls for an increase in sensitivity for detecting emergent qualities. One the other hand, it concludes that emergence of new qualities can be a source of a city's unique characteristics, setting it apart from other cities that try to re-implement 'best practices' from elsewhere.

Keywords Complex urban systems · Emergence of new qualities · Limits to planning · Urban development with unpredictable new qualities · Central and Eastern European cities

This chapter seeks to explain the challenge of, increase the sensitivity for, and offer an approach for taking emergent, i.e. unplanned and in principle unpredictable, qualities in urban development into consideration. I argue that emergent qualities in complex urban systems can be of significant importance to urban development, and I suggest that, in order to learn about emergent qualities, citizens can be asked not about their wishes but about what they observe that appears new in the city. This conclusion is drawn upon a review of emergent urban qualities and

---

C. Walloth (✉)
University of Duisburg Essen, Institute of City Planning and Urban Design, Universitätsstraße 15, V15 S04 D03, 45141 Essen, Germany
e-mail: christian_walloth@complexurbansystems.net

C. Walloth et al. (eds.), *Understanding Complex Urban Systems: Multidisciplinary Approaches to Modeling*, Understanding Complex Systems,
DOI: 10.1007/978-3-319-02996-2_8,

Fig. 1 The particular urbanity of Lviv's old market square emerges from the constellation of actually existing physical and cultural qualities (photo by the author, 2009)

the observation that urban planning might overlook the possibility of emergent qualities and their potentially significant impact in urban systems.

The arguments are illustrated using examples from second-tier (i.e. non-capital) cities in Central and Eastern Europe (CEE)—"over the last two decades … a fascinating laboratory" (Ogden 2011, p. xvi). They have, first, developed different material and social structures as compared to their Western counterparts which, hence, allows for comparison between developments and, second, are still in transition, which enables the observation of many concurrent, ongoing developments. "[I]nherited urban structures" (Turnok 1989, p. 151)—from the medieval core to the communist blocks of flats—influence today's social and, in turn, material form of the CEE second-tier city. Other influences are provided by national and supra-national processes or conditions, the cities being "part of global (or at least European) networks or economies" (van Kempen et al. 2005, p. 254). With these many influences—interdependent processes and structures—acting from within and outside, CEE second-tier cities certainly belong to the type of complex urban systems that surprise us with unpredicted, emergent developments (Fig. 1).

The complex urban system, as any complex system, is unpredictable in that it neither behaves like a mechanism (causally) nor like a statistically random population (probabilistically) (cf. Weaver 1948, p. 583). If such unpredictability, in which the emergence of ever new qualities plays a key role, is not to lead us to question the effectiveness of urban planning per se, it must at least call for urban planning to widen its scope towards the handling of emergent qualities. Approaches might aim at leveraging synergies with positively perceived emerging qualities, or strategies of coping with undesired ones. This article calls for a conscious handling of emergence in urban systems.

1 Known System Approaches Cannot Model Emergence

It seems suggestive here to apply systems approaches to improve the understanding of—and thus, the effectiveness of planning in—complex urban systems. Indeed, system approaches can be of great help in planning practice as demonstrated by, e.g., Cole (2007). Well-known urban systems approaches have been introduced by Forrester (1969) and Vester (1988), where Vester's biocybernetic approach relates to and extends the system dynamics approach by Forrester. Besides, agent-based modeling is frequently applied to understand the behavior of complex systems, e.g., spatial arrangements of people, revitalization of old towns, or the appearance of shopping malls (see also Atun 2014 and Gebetsroither-Geringer 2014, both in this volume).

Forrester's and Vester's approaches make use of system diagrams in which (directed, weighted) relations between elements of the system under consideration are depicted. They have been applied and discussed in several studies (e.g., Vester and Hesler 1980; Cole 2007; Grosskurth 2007; Walz et al. 2007) and handbooks (e.g., Scholtz and Tietje 2002) with regard to urban and regional planning. However, since both approaches are rooted in the research of the 1960s and 1970s, they mainly highlight feedback and relational characteristics while omitting more recent findings about urban systems, such as self-organization (see, e.g., Portugali 1997) and emergence (see, e.g., Deppisch and Schaerffer 2010, p. 25–26) (Fig. 2).

Self-organization can be understood by modeling the interactions between agents according to a given set of rules. In agent-based modeling, the modeling results are within the limits of what can be expected as outcomes of the interaction of agents that make up the model. If the agents represent people who move between neighborhoods, who give birth and die, who start commercial activities and buy, then the model could, e.g., simulate various distributions of residential and commercial areas, and the evolution of these over time. This evolution can be

Fig. 2 Who could have imagined and modeled the emergence of this type of homy coffee shops in Poznań before they first appeared? (photo by the author, 2009)

based on initial and learned behavior of the agents, and on random changes (fluctuations, mutations) of behavior.

These approaches can yield information about feedback dynamics and self-organization, but not about emergence, which is the coming into being of new quality. Emergence is what transcends the *a priori* known space of qualities. It leads to "real novelty … much that was not foreseeable, at least not for human knowledge" (Popper and Eccles 1977, p. 16). The outcome of the existing system approaches is limited to such qualities which are within the *a priori* known space of qualities. If, as mentioned above, in agent-based modeling, a spatial distribution of agents with some behavior is the starting point, then the output will be a spatial distribution of agents with some behavior. The simulation tells us something within the limits of the known qualities only, in this case the spatial distribution after a certain time. Emergence, however, cannot be explained by any of these approaches that operate with known units, since emergence affords a new quality. This quality is new in the sense that it cannot be predicted from knowing the parts of the system out of which the new quality emerges (c.f. Popper and Eccles 1977, p. 22 ff) (Fig. 3).

2 Emergent Qualities are Unpredictable—Yet They May Have Significant Influence on Urban Development

2.1 Unpredicted Love Locks

Emergent qualities cannot be predicted, yet they may interfere with or influence urban development strategies. An example of an unpredicted and unplanned emergent development is that of the Hungarian city of Pécs as a "City of Lovers" by the symbolism of its "Love Locks", as described by Hammond (2010):

> Thousands of metal padlocks have transformed this East-Central European city. […] Often inscribed with initials and dedications, the locks are rumored to represent young paramours' hopes for the longevity of their passions. (p. 181)

and further:

> What the locks do offer is a radical aesthetics of interruption, which, even in their recuperation, are an indelible trace of how cities emerge through oppositionality. … [T]hey are the telling detail which has the potential to unravel the image as a whole. (p. 192)

Now that it has emerged, this quality impacts on the image of the city, urban life, the touristic development and, through the latter, the wealth of the city and its citizens—it has a tremendous impact on the city. The city of Pécs has integrated the Love Locks into its cultural and marketing image, and "travel websites […] describe Pécs as a 'City of Lovers'" (p. 181). But who could have predicted that from somebody adding a first padlock onto a gate, a bottom-up process would lead to the emergence of new cultural and image aspects of the city, of new possibilities of urban marketing, tourist attraction, historical narration?

Fig. 3 Love Locks (here in Riga) appeared in many cities over the past two decades. What will emerge next? (photo courtesy of Irina Minciuna, 2007)

Emergent qualities might come as a positive (as above) or negative surprise, or urban actors might wish for them to appear in order to make urban development strategies work. The latter is a challenge, since emergence cannot be planned—we cannot know the new quality before it appears and, hence, we cannot devise the qualities and their required quantities which could lead to an emergent quality (Fig. 4).

Hence, despite the potentially great influence on the development of a city, urban planning is not able to design and implement emergent qualities.

2.2 Sought-After New Knowledge Culture

Such inability to plan emergence may pose a challenge, e.g., in case a new, superseding cultural understanding is required for cross-border, cross-cultural urban development. This can be illustrated using the example of a superseding cultural understanding aimed at in a German-Polish joint-venture between the universities of Poznań in western Poland and Frankfurt an der Oder in eastern Germany. The joint-venture, the Collegium Polonicum located in Słubice, Frankfurt an der Oder's twin town just across the Oder river, is meant to be a building stone in the region's post-industrial and post-communist transformation into the knowledge society (Fichter-Wolf and Knorr-Siedow 2008). A "knowledge utilization culture" was supposed to emerge in order to stimulate this transformation, but was hindered by "institutional 'misfits'" that could not be overcome (pp. 44–45). While "*new* knowledge cultures" (p. 51, emphasis by the original authors) would have been the necessary, emergent more-than-the-sum-of-its-parts "elements of a new quality" (p. 51), only "convergence" is practiced, with "decisions … taken by consensus … in order to minimize the dangers of culturally based conflicts" (pp. 47–48). The achieved "consensus" resorts to the lowest

Fig. 4 Lack of emergence of a proper pedestrian zone in the city of Perm (photo by the author, 2012)

common denominator and a subset of the qualities already brought in from both sides.

These examples show how emergent qualities such as a new knowledge culture could not be planned but may only "*happen* in the intermediary [cultural] spaces and niches" (pp. 50–51, my emphasis). As Fichter-Wolf and Knorr-Siedow put it: "[T]he development of a … knowledge culture … is rarely a large design" (p. 51).

Emergence, hence, is a challenge to urban planning. It cannot be predicted, and it cannot be designed. It might be expected and will not happen. It may yield qualities that fit with current urban development plans or not, that might have a significant impact on urban development or not. Its fundamental unpredictability may pose a major challenge that might never be overcome. Questions important for urban planning practice are, e.g., how to deal with emergence once it happens, and how to not rely on it since it may not happen as desired. Subsequently, I will discuss the aspect of the sensitivity of urban development to initial conditions, which, in conjunction with emergence, offers insights into the limits of transferability of learnings and so-called 'best-practices' from one urban system to another.

3 Urban Planning Tends to Overlook the Possibility of Emergence

A central element of the scientific scheme of conjecture and testing of hypotheses is the experiment that can be repeated. Considering urban interventions as experiments in urban space, the application of 'best practices' in different cities at different times rests on the same premise of result repeatability.

However, in environments such as urban systems, where emergent qualities lead to successive qualitative changes of the setting, the repetition of experiments

Fig. 5 High-rise buildings in Gdynia—only one contribution to emergent urban quality (photo by the author, 2012)

does not guarantee the same results. Similar material structures, when combined with different socio-cultural settings, might lead to different urban developments and give rise to entirely different emergent qualities. An apparently similar urban setting might develop radically different urban functions (social, cultural, economic etc.) in different cities at different times.

3.1 Ambiguous Developments of Residential High-Rises

This argument may be illustrated using the development of residential high-rise buildings in CEE versus western Europe and the US. Large-scale housing was built both in former communist cities and Western cities. The learning from the Western cities is clear: Large-scale housing estates are associated with crime and people dependent on welfare. The 1960's high-rises of Chicago's Robert Taylor Homes and Cabrini-Green settlements became social hot spots and were eventually torn down. The high-rise neighborhood of Köln Chorweiler, finished in 1972, is notorious for its social problems, high unemployment rate and large share of migrants. And the banlieues (suburbs) of French cities are similarly infamous for their social problems.

Quite in contrast, in CEE cities a totally different use-pattern has emerged (Fig. 5). More than twenty years after the fall of communism, the *paneláky* (blocks of flats in colloquial Czech) "are, for the most part, still stable social areas" (Grossmann and Steinführer 2011, p. 24). The apartments in the communist blocks of flats are still "sold and bought at good prices on the free market. Their social mixture determines these estates not to be regarded as derogatory in terms of social prestige, as they more typically are in Western Europe" (Vais 2009, writing about the situation in the city of Cluj, citing Murie et al. 2005).

3.2 Overlooked Differences

Such emerging differentiation of urban trajectories is often overlooked in urban research and development. The treatment of the post-communist CEE city provides an example of how, on the one hand, urban research and development tries to apply categories known from the history of Western cities to different urban systems while, on the other hand, the actual developments diverge significantly. For example, Kunzmann (2006) notes:

> In gentrified inner cities, attractive suburban communities and in leisure regions the nouveau-riche upper class is showing its wealth. In contrast, those who did not succeed in benefiting from the new market economy are forced to remain at the urban fringe of agglomerations, in the run-down pre-fabricated social housing schemes where Western supermarkets and fast food chains absorb their small incomes. (p. 21)

However, the transformation of the Central European city towards such a typical western place is now highly disputed, because different initial settings have yielded different paths of development (Fig. 6). Pichler-Milanovič (2005) observes:

> In the 1990s it was rather assumed that transition [...] would project cities in Central and Eastern Europe rather uniformly along a linear trajectory, which would result in their convergence [...] with those [cities] in Western Europe. Such thinking, however, was not only naïve in the light of subsequent reality, but was often based on a lack of understanding [...]. [C]ities in Central and Eastern Europe are 'path dependent' on their pre-socialist as well as their socialist-period legacies.

These observations exemplify how urban systems develop different qualities over time. In other words, there are qualities in urban systems which do not show the tendency to equifinality but, in contrary, to differentiation—due to different and unique initial situations and different successions of emergent qualities over time. Whenever something new comes in, like residential high-rises, the new situation in turn opens up further new, and potentially unique, possibilities for emergent development (cf. Popper and Eccles 1977, p. 30). This results in ever new possible emergent qualities that differ from city to city—something which might be overlooked in urban planning in favor of applying known categories and 'best practice' schemes of intervention.

4 Emergent Qualities as a Challenge to Urban Planning

A consequence of the above considerations is that what works in one city might not work in another. And what emerges in one city might not emerge in another city although the emergent quality would be welcome or even desired. For example, copying the physical elements of a successful neighborhood somewhere (and, to a certain extent, also getting the 'right' people in, e.g., the now fashionable 'creative class') will be no guarantee for the emergence of a vivid and thriving 'urban' neighborhood elsewhere. Something entirely different might

Fig. 6 Different patterns have emerged in CEE cities than in the West. Run-down inner-city residential buildings from the end of the nineteenth century in Wrocław (**a**) and kiosks in place of supermarkets in Iaşi (**b**) (photos by the author, 2012)

happen, and what emerges may be undesired. In any case, emergent qualities come as a surprise.

Yet, emergent qualities may have a significant impact on urban development. On the one hand, the emergence of stable social environments in areas of residential high-rises or the emergence of tourism-generating Love Locks, though unpredicted and unplanned, might be welcome. On the other hand, there are emergent qualities such as a social decline of a residential high-rise neighborhood, which urban development might want to stop.

4.1 Planning with Emergence

What can be the role of purposeful urban planning in urban systems that develop emergent qualities? Acknowledging the unpredictability of emergent qualities, urban planning can neither design them, nor prevent them before they appear: Emergent qualities may or may not arise from certain constellations of existing urban qualities.

Thus, one role of urban planning might be in creating or preventing certain constellations which might allow new qualities to emerge or might prevent others from emerging, as it was (unsuccessfully) tried in the case of the Collegium Polonicum, referred to above. Such 'planning' activity might seek to prepare the ground either by implementing interventions conjectured as triggers for emergence or by relieving constraints conjectured to choke emergence. Creating physical space for knowledge sharing could be one such intervention that might (!) lead to a desired emergent quality. The enforcement of padlock-free inner-city fences would have choked the emergence of the “City of Lovers”.

Another role of urban planning lies in the accommodation of newly emerged qualities, i.e. with change that cannot be foreseen. Such change may either be seen as positive or negative in the light of urban development strategies. In the first, positive case, synergies with desired outcomes of urban planning might be achievable. In the second, negative case, adaptation might be required to incorporate, though not to support, and to cope with the newly emerged quality.

4.2 Detecting Emergence

A primary precondition for dealing with emergence is the increased awareness and sensitivity of urban research and urban planning for emergent qualities and their potentially significant impact on urban development.

Since emergent qualities cannot be predicted, an early detection is of paramount importance. By recognizing newly emergent qualities as early as possible, it might be possible to deal with them—in whatever way—before they become dominant. In analogy to earth quake detection, urban planning would require a seismometer for detecting emergent qualities in the city.

One obvious possibility could be to involve citizens as reporters or ‘trend scouts’ for new things they observe coming up. Hence, citizens would be early ‘seismic’ indicators for emerging qualities—they might not be able to predict, but they can provide a feeling for what has most recently emerged and might subsequently grow to importance. To a certain extent, they might even anticipate what may come, i.e. they generate emergent thoughts, which would be, in a sense, a simulation of emergence in an interactive setting.

5 Summary and Conclusion

Emergence, i.e. unpredictable change, can be both: curse and blessing of urban planning efforts. What is important in urban planning is to be aware of the possibility of being surprised by emergent qualities that might have a significant influence on the development of the city. Surprise might stem from the fundamental unpredictability of emergent qualities as well as from different effects of comparable interventions in different cities. Such different reactions to the same interventions are caused by each city’s unique situation, formed by the succession of emergent qualities over the city’s history.

Opportunities may arise from emergent qualities, like the "City of Lovers" image of Pécs. For such opportunities to happen, certain forms of emergence need to be welcomed. Emergence might even be stimulated by trying out different actions in the city to set impulses and test if they contribute to emergent effects. However, some emergent qualities might not fit the urban development goals and we might want to detect them before they become dominant. In order to detect what emerges, citizens might be involved as 'reporters of change'.

The concept of emergence in urban systems is important, because it opens our eyes for the yet unknown and should make us wary of simply copying would-be 'best practice'. It also leads us to embrace uniqueness. The source of a city's potential for uniqueness may lie in emergent qualities rather than in copying what has been done elsewhere before.

Acknowledgements I am indebted especially to Arun Jain and J. Alexander Schmidt for their thorough review of the first version of this contribution. They encouraged me to turn a mere essay into a more rigorously written article.

References

Atun, F.: Understanding effects of complexity in cities during disaster. In: Walloth, C., Gurr, J.M., Schmidt, J.A. (eds.) Understanding Complex Urban Systems: Multidisciplinary Approaches to Modeling, pp. 51–65. Springer, Cham (2014)

Cole, A., Allen, W., Kilvington, M., Fenemor, A.: Participatory modeling with an influence matrix and the calculation of whole-of-system sustainability indicators. Int. J. Sustain. Dev. **10**(4), 382–401 (2007)

Deppisch, S., Schaerffer, M.: Given the complexity of large cities, can urban resilience be attained at all? German Annual of Spatial Research and Policy, 2010, pp. 25–33. Springer, Berlin (2010)

Fichter-Wolf, H., Knorr-Sidow, T.: Transformation processes and cross-border cultures—a milieu-sensitive approach toward fractures of decline and growth. In: Nowak, M., Nowosielski, M. (eds.) Declining Cities/Developing Cities: Polish and German Perspectives, pp. 31–56. Instytut Zachodni, Poznan (2008)

Forrester, J.: Urban Dynamics. Productivity Press, Portland (1969)

Gebetsroither-Geringer, E.: Multimethod modeling and simulation supporting urban planning decisions. In: Walloth, C., Gurr, J.M., Schmidt, J.A. (eds.) Understanding Complex Urban Systems: Multidisciplinary Approaches to Modeling, pp. 13–27. Springer, Cham (2014)

Grosskurth, J.: Ambition and reality in modeling: a case study on public planning for regional sustainability. Sustain. Sci. Pract. Policy 3(1), 3–11 (2007)

Grossmann, K., Steinführer, A.: Residential change: conceptualization, methodological challenges and research design. In: Haase, A., Steinführer, A., Kabisch, S., Grossmannn, K., Hall, R. (eds.) Residential Change and Demographic Challenge. The Inner City of East Central Europe in the 21st Century, pp. 17–37. Ashgate, Farnham and Burlington (2011)

Hammond, C.: Renegade ornament and the image of the post-socialist city. In: Kovacs, T. (ed.) Halb-Vergangenheit. Städtische Räume und urbane Lebenswelten vor und nach 1989. Lukas-Verlag, Berlin (2010)

Kunzmann, K.: Spatial development and territorial cohesion in Europe. In: Altrock, U., Güntner, S., Huning, S., Peters, D. (eds.) Spatial Planning and Urban Development in the New EU Member States, pp. 19–30. Ashgate, Aldershot (2006)

Murie, A., Tosics, I., Aalbers, M., Sendi, R. Cernic-Mali, B.: Privatisation and After. In: van Kempen, R., Dekker, K., Hall, S., Tosics, I. (eds.) Restructuring Large Housing Estates in Europe, pp. 85–104. The Policy Press, Bristol (2005)
Ogden, P.: Living in the European city: demographic change and residential patterns. In: Haase, A., Steinführer, A., Kabisch, S., Grossmannn, K., Hall, R. (eds.) Residential Change and Demographic Challenge. The Inner City of East Central Europe in the 21st Century, pp. xv–xvii. Ashgate, Farnham (2011)
Pichler-Milanovic, N.: Urban Development in Central and Eastern Europe: From Transition to »Creative« Competition? 41st ISoCaRP Congress (2005)
Popper, K., Eccles, J.: The Self and Its Brain. Springer, Berlin (1977)
Portugali, J.: Self-organizing cities. Futures **29**(4/5), 353–380 (1997)
Scholz, R., Tietje, O.: Embedded Case Study Methods. Sage, Thousand Oaks (2002)
Turnok, D.: The Human Geography of Eastern Europe. Routledge, London (1989)
Vais, D.: From House to "Residence". Peripheral Growth in Post-Socialist Cluj (Romania). Sociologia LIV (1), 75–78 (2009)
van Kempen, R., Vermeulen, M., Baan, A.: Urban issues and urban policies in the new EU countries: synthesis and conclusions. In: van Kempen, R., Vermeulen, M., Baan, A. (eds.) Urban Issues and Urban Policies in the New EU Countries, pp. 253–268. Ashgate, Aldershot (2005)
Vester, F., Hesler, A.V.: Sensitivitätsmodell. Umlandverband, Frankfurt (1980)
Vester, F.: The biocybernetic approach as a basis for planning our environment. Syst. Pract. **1**(4), 399–413 (1988)
Walz, A., Lardelli, C., Behrendt, H., Grêt-Regamey, A., Lundström, C., Kytzia, S., Bebi, P.: Participatory scenario analysis for integrated regional modelling. Landscape Urban Plan. **81**(1), 114–131 (2007)
Weaver, W.: Science and complexity. Am. Sci. **36**(4), 536–544 (1948)

'Urban Complexity' from a Literary and Cultural Studies Perspective: Key Cultural Dimensions and the Challenges of 'Modeling'

Jens Martin Gurr

Abstract This contribution sketches a literary and cultural studies approach to urban complexity. It briefly comments on the neglect of cultural phenomena in much urban modeling, discusses key characteristics of urban complexity from a literary and cultural studies perspective and relates these to technical or mathematical notions of complexity. After engaging with the need for reduction and compression in modeling and with the resulting limitations of urban models, a more specifically 'literary' section discusses simultaneity as a central characteristic of urban complexity and as a challenge to narrative representation and points out a number of strategies in the 'modeling' of urban complexity in literature. The article thus sets out to show what literary studies can contribute to truly interdisciplinary work on urban complexity and urban systems, arguing that literary studies specifically bring back into the discussion those features of urban complexity that resist quantitative modeling but that are nonetheless crucial to a differentiated understanding of the functioning of urban systems.

Keywords Complex urban systems · Limits to planning · Limits to quantitative modeling · Literary strategies · Narrative modeling of complexity · Narrative representation · Qualitative modeling · Simultaneity

1 Introduction

This contribution attempts to sketch a literary and cultural studies approach to urban complexity: After briefly commenting on what I perceive as a neglect of cultural phenomena in much urban modeling, I will discuss key characteristics of

J. M. Gurr (✉)
Department of Anglophone Studies, Joint Center "Urban Systems",
University of Duisburg-Essen, Universitätsstraße 12, 45141 Essen, Germany
e-mail: jens.gurr@uni-due.de

C. Walloth et al. (eds.), *Understanding Complex Urban Systems: Multidisciplinary Approaches to Modeling*, Understanding Complex Systems,
DOI: 10.1007/978-3-319-02996-2_9,

urban complexity from a literary and cultural studies perspective and will relate these to technical or mathematical notions of complexity. My third section engages with the need for reduction and compression in modeling and with the resulting limitations of urban models before discussing the fundamentally different status of the model in urban complexity studies on the one hand and in literary and cultural studies on the other hand. In a more specifically 'literary' section, I discuss simultaneity as a central characteristic of urban complexity; I here outline the challenge it poses to literary representation and point out a number of literary strategies in the 'modeling' of urban complexity in cultural production and specifically in literature.

The essay thus sets out to show how literary and cultural studies engage with urban complexity and what such an engagement can contribute to truly interdisciplinary work on urban complexity and urban systems. I will argue that literary and cultural studies specifically bring back into the discussion those features of urban complexity that resist modelling but that are nonetheless crucial to a differentiated understanding of the functioning of urban systems.

2 The Neglect of 'Culture' in Urban Complexity Research

In his masterful study *Die komplexe Stadt: Orientierungen im urbanen Labyrinth* (2009), one of the most recent and ambitious attempts to provide urban studies with an integrating research agenda, Frank Eckardt goes so far as to propose complexity as the key characteristic of the city and calls for a transdisciplinary research programme organized around the integrating paradigm of 'complexity'. However, Eckardt then largely pursues a sociological programme.

It seems that, despite much talk of multi-, inter- or transdisciplinary research in (urban) complexity, the perspective of culture is curiously absent.[1] More specifically, while the consideration of actors and groups of actors and their behaviour as an important dimension of 'culture' as studied by the social sciences is clearly central to modeling endeavours, patterns of symbolic representation and patterns of perception and interpretation as less tangible elements of the city largely resist modeling and are a conspicuous absence in models of urban complexity. A few prominent examples may suffice here: The "Preface" to one of the most ambitious collections on the topic, Albeverio et al.'s *The Dynamics of Complex Urban Systems*, in its plea for a "fruitful collaboration between natural science" and "regional science" mentions "physics, mathematics, computer science, biology, ..." (omission original) on the side of "natural science" and "architecture, geography, city plannings [sic], economics, sociology, ..." (omission original) on

[1] The same is true of the roles and functions of urban culture in urban systems. For discussions of the functions of urban culture and the contribution of urban cultural studies to transdisciplinary urban research, cf. for instance Butler and Gurr 2014, forthcoming, Gurr 2011a, b, Gurr and Butler 2011, 2012.

the side of "regional science" (2008, p. v)—and the volume, comprehensive as it is, does not contain anything even remotely from the field of cultural studies. Even in the exhaustive and masterful 2009 *Encyclopedia of Complexity and Systems Science*, Michael Batty's state-of-the-art contribution on "Cities as Complex Systems: Scaling, Interaction, Networks, Dynamics and Urban Morphologies" (pp. 1041–1071) does not even mention 'culture', nor do concepts like 'individuality' play any role. Similarly, "Springer Complexity" according to the description for what is arguably the leading and most prominent book series in the field, is said to "cut across all traditional disciplines of the natural and life sciences, engineering, economics, medicine, neuroscience, social and computer science" (Portugali 2012, front matter). Moreover, the outline for Springer's "Understanding Complex Systems" (UCS) series, in which this volume appears, states the aim of the programme as follows: "UCS is explicitly transdisciplinary", its first "main goal" being "to elaborate the concepts, methods and tools of complex systems at all levels of description and in all scientific fields, especially newly emerging areas within the life, social, behavioural, economic, neuro- and cognitive sciences (and derivatives thereof)" (Portugali 2011, front matter). The mental, non-institutional dimension of culture in the form of symbolically mediated patterns of perception and interpretation of human environments, it seems, hardly features in discussions of urban complexity.

3 Characteristics of (Urban) Complexity: Urban Systems Research and the Perspective of Literary and Cultural Studies

Many of the characteristics of urban complexity frequently discussed in research on complex urban systems[2] are those that are also of interest to literary and cultural studies. Portugali (2006) conveniently defines the complexity of the city thus: "a very large number of interacting parts, linked by a complex network of feedback and feedforward loops, within a system that is open to and, thus part of, its environment" (p. 657; cf. also Portugali 2011, p. 232). Further characteristics of complexity frequently discussed include self-organization, emergence, nonlinearity, phase transitions, density, mobility (as one cause of change over time and as the occasion for increased interaction and mixing), ethnic and cultural multiplicity, heterogeneity and hybridity, violence, conflicts over the use of space, intersections of technology and virtual spaces with physical and experienced spaces,

[2] For enlightening explicit or implicit definitions of urban complexity and its central features, cf. for instance Batty 2009, passim; Eckardt 2009, passim; Mainzer 2007, p. 374 et passim; Portugali 2000, 2006, pp. 652–657 et passim; Portugali 2011, p. 232ff. et passim; Portugali 2012, p. 4f).

overlapping and intersecting spatial scales—from the local to the global—and their interdependencies, as well as complex interferences, interdependencies or intersections in the interaction between multiple players, intentions, or force-fields, which together make urban systems prime examples of translocal networks of complex relationships, connections and interdependencies subject to rapid change over time. Moreover, as already indicated in Portugali's working definition, the city is of course an open or dissipative rather than a closed system (in the technical sense): It exchanges goods, energy, information, people, money etc. with its environment. All these characteristics, it seems, are more or less part and parcel also of urban simulation models in the more technically oriented disciplines concerned with urban modeling from a complex systems perspective.

However, the 'softer', less easily quantified and modeled characteristics of urban complexity are no less central to the 'urban experience'. In this vein, in addition to the 'usual suspects' such as "hierarchy and emergence, non-linearity, asymmetry, number of relationships, number of parts", Mainzer (2007, p. 374) also lists the following features: "values and beliefs, people, interests, notions and perceptions" (they appear in a visualization, hence in no particular order—the order is mine). Additionally, while the notion of a system's 'history'—in the sense that previous developments have an impact on the present and future course of the system—is central to urban modeling (if only in the sense that past developments can be extrapolated for predictive purposes), one specific aspect of a city's history is of particular importance to an understanding of urban systems from the perspective of literary and cultural studies: This is the notion of the city as a palimpsest, a form of layered spatialized memory.[3] Individual and collective memory and the way it is physically manifested in and evoked by the built environment is a prime concern in many literary texts (for a detailed discussion of a specific key text, cf. Gurr 2014, forthcoming).

Arguably the central challenge to a literary 'modeling', however, lies in the multiplicity of sense impressions and the resulting sensual overload and semiotic 'overkill' as a result of a multiplicity of sign systems and ceaseless semiosis. In this vein, Georg Simmel in his influential 1903 essay "The Metropolis and Mental Life", one of the crucial texts in the early phase of urban studies, defines "the psychological conditions which the metropolis creates" as marked by the "*intensification of nervous stimulation* which results from the swift and uninterrupted change of outer and inner stimuli… the rapid crowding of changing images, the sharp discontinuity in the grasp of a single glance, and the unexpectedness of onrushing impressions" (p. 13, italics original).

In his recent study *Complexity, Cognition and the City*, in a remarkably similar way, Portugali speaks of

[3] For various aspects of the the notion of the city as a palimpsest, cf. Assmann 2009; Harvey 1989, p. 66; Hassenpflug 2006, 2011; Huyssen 2003; Martindale 1995; Sharpe and Wallock 1987, p. 9; Suttles 1984. For the palimpsest in literary studies, cf. especially Dillon 2007.

> the really complex situation that concerns individuals under a bombardment of information, that is, under a multiplicity of messages from a multiplicity of sources and of all kinds. This is typical of the dynamics of cities: every agent operating in the city is continually subject to a multiplicity of messages in the form of views, noises, smells etc. In order to behave and survive, the agent… must make sense of all those signals and messages (Portugali 2011, pp. 232f).

As I will argue in Sect. 5., it is precisely the representation of simultaneity that poses the greatest challenge in the narrative representation of urban complexity.

4 Urban Complexity in Technical and Literary 'Models': Information, Compression, the Limits of Modeling—and the Status of the Model

In two chapters jointly authored with Hermann Haken that largely go back to an earlier joint essay, Portugali elaborates on the amount of information contained in the physical structure of the city.[4] In the sub-chapter tellingly entitled "How Many Bits to the Face of the City" (pp. 179–186), the authors argue: "From the point of view of information theory, the face of the city is a message. As a message it conveys and transmits different quantities of Shannonian information" (Haken and Portugali 2003, p. 403). Conceived in terms of informational complexity, it is obvious that the city—even if we just take its physical shape and ignore for the moment the interaction of millions of humans each with their own thoughts, hopes, anxieties etc.—contains a virtually endless amount of information.

As far as the 'measurement' of complexity from a literary and cultural studies perspective is concerned, it is evident that most mathematical or technical measures are hardly helpful: Neither algorithmic or Kolmogorov complexity (i.e., the length of the minimal programme that reproduces a given sequence, for a discussion, cf. Li and Vitányi 1993), nor Bennett's notion of logical depth (essentially a measure of the time required for a given bit string to be computed and reproduced), nor effective complexity in the sense of Gell-Mann (for a discussion, cf. for instance Gell-Mann 1995a, b), nor the measure of informational entropy, nor the question of calculability as in P- or NP-problems, nor dynamic complexity are really helpful here. It seems that, when descriptions of urban systems and the specific forms of complexity they exhibit are concerned, it does not really matter whether we are looking at Kolmogorov complexity or effective complexity, or logical depth—the multiple interdependencies, overlapping scales, forms of self-organization etc., yield an astonishing degree of complexity by any definition. However, this complexity is not beyond comprehension or calculability: "The significant achievement of complexity theories is to show that even [under such

[4] Cf. Portugali 2011, Chap. 8: "Shannonian Information and the City" (pp. 167–187) and Chap. 9: "Semantic Information and the City" (pp. 187–210); cf. also Haken and Portugali 2003.

complex conditions] a scientific approach is possible." The solution, Portugali argues, lies in what he refers to as "information compression" (2011, pp. 231–233).

Compression and strategic 'reduction' of complexity are of course part and parcel of any modeling process. The crucial task, of course, is to decide what can safely be left out or abstracted so as not to distort the overall picture. This will naturally depend on what the model is supposed to achieve—what, in other words, is of interest and what can be left out? In this context, Cohen and Stewart (1995, p. 410) aptly remark:

> Mathematical descriptions of nature are not fundamental truths about the world, but models. There are good models and bad models and indifferent models, and what models you use depends on the purposes for which you use it and the range of phenomena which you want to understand... Reductionist rhetoric... claims a degree of correspondence between deep underlying rules and reality that is never justified by any actual calculation or experiment.

Lefebvre similarly comments on the need for reduction in dealing with "complexity" but also on the inherent dangers:

> Reduction is a scientific procedure designed to deal with complexity and chaos of brute observations. This kind of simplification is necessary at first, but it must be quickly followed by the gradual restoration of what has thus been temporarily set aside for the sake of analysis. Otherwise a methodological necessity may become a servitude, and the legitimate operation of reduction may be transformed into the abuse of reductionism (1991, pp. 105f).

In a related vein, David Byrne in an excellent overview on *Complexity Theory and the Social Sciences* aptly remarks that complexity science is "clearly quantitative" in its aims and methods, and appropriately points out three more fundamental caveats and limits to quantitative analysis and modeling:

> 1. The limits to formalisation of any mathematical system established by Gödel
> 2. The limits to capacity of measurement central to deterministic chaos; and
> 3. The working limits for the expression of mathematical formalism derived from the non-linearity of the real systems with which chaos/complexity is concerned. (1998, p. 5; for an excellent discussion of these caveats, cf. Byrne 1998, pp. 54–71)

More specifically and with a view to practical limitations in the modeling of urban systems, Portugali (2012) in an enlightening state-of-the-art article on "Complexity Theories of Cities: Achievements, Criticism and Potentials", comments on a key problem in many contemporary applications of complexity theory in urban modeling:

> There is nothing wrong... in sophisticated simulation models crunching huge quantities of data by means of fast computers. What's wrong is... that simulation models originally designed as media by which to study phenomena of complexity and self-organization become the message itself. (p. 52)

As a result, "practitioners of urban simulation models tend to overlook the non-quantifiable urban phenomena" (p. 52).[5]

In his "Introduction" to the same volume, Portugali (2012, p. 4) attempts to account for this loss by commenting on the research motivation of scholars in CTC (complexity theories of cities): "Some are physicists for whom cities is [sic] just another source for quantitative data with which to test their models, while others are urbanists who see CTC as the new and more sophisticated generation of the 'old' quantitative approach to cities. By so doing they overlook the qualitative message of CTC" (p. 4).[6] He then incisively asks: "But what about the uniqueness of cities—of the properties that differentiate them from material and organic entities, how do these [relate] to their complexity and dynamics?" (p. 4). Thus, what is missing, according to Portugali, is only the analysis of what distinguishes cities generally from other complex systems; he does not appear to be interested in the arguably more important question what makes an individual city unique—as in studies on the "intrinsic logic ['Eigenlogik'] of cities" (cf. Berking and Löw 2008)—, let alone in the uniqueness of individuals and their response to the city.

This, I argue, is where literary and cultural studies come in: For although it is of course possible to include certain 'subjective' features into a model (for instance by including group-specific cultural preferences etc.), what is individual, unique, historically and personally specific and not reducible to an underlying pattern is what disappears in abstracting from the individual and in the aggregation of preferences, needs, desires, hopes, fears into an equation—and this is what literary and cultural studies find most interesting.

What is the consequence of this type of reduction? While Rolf Lindner (2008, p. 92) has argued that "[t]he city of sociologists... is frequently a non-sensual place, a city one does not hear, smell, taste, more precisely, a Non-Place",[7] I would argue that this is even more true of the 'city of modelers and complexity theorists'.

---

[5] Similarly, Portugali (2012, p. 54) argues that "[q]ualitative urban phenomena do not lend themselves to quantitative-statistical analysis and thus are of little interest to mainstream CTC".

[6] Portugali (2011, p. 227) somewhat schematically accounts for this by commenting on the different methodologies of the natural as opposed to the social sciences: "The methodological tools of the 'hard' sciences are reductionism, mathematical formalism, statistical analysis and explanation, while those of the 'soft' humanities and social theory are the exact opposite: anti-reductionism, understanding in place of explanation, and hermeneutics in place of analysis." Like Portugali, Mainzer (2007, p. 12) calls for a recognition of the qualitative features of a system and argues that complexity science can function as a connection: "Contrary to any reductionistic kind of naturalism and physicalism we recognize the characteristic intentional features of human societies. Thus the complex system approach may be a method of bridging the gap between the natural sciences and the humanities that was criticized in Snow's famous 'two cultures'." For an enlightening if frequently schematic problematization of "the two cultures" in the context of complexity theories and cities, cf. Portugali 2011 (pp. 9–52); cf. also Stephen Read (2012) and several other contributions in Portugali et al. 2012.

[7] "Die Stadt der Soziologen hingegen ist für gewöhnlich ein unsinnlicher Ort, eine Stadt, die man nicht hört, nicht riecht, nicht schmeckt, genau genommen ein Nicht-Ort." (Marc Augé's notion of the non-place (1995) is not quoted here, but clearly implied, it seems).

A further key issue that needs to be addressed in comparative discussions of 'urban modeling' in technical complexity research and in literary studies is the fundamentally different status of the 'model' in both fields: While in technical urban complexity research, the model is the *result* of scientific endeavour, in literary and cultural studies, the literary text functions as the 'model' and is thus the object of study rather than the result of the scholar's own work.

5 Literary 'Models' of Urban Complexity: The Challenge of Simultaneity and Selected Strategies

From the perspective of literary and cultural studies, the question is how the complexity of the urban text is 'modeled' in literature (and other media, though I do not discuss them here).[8] Despite the universally diagnosed importance of complexity as a key characteristic of the urban and despite the widely perceived affinity between the city and the novel (and film)—and a look at innumerable 'urban' poems and narrative fictions confirms that the issue is indeed central to 'urban literature'—the specific issue of how literary texts represent urban complexity has received very limited explicit scholarly attention. In most of the innumerable studies on urban imaginaries in literature and film, though occasionally implied, complexity—let alone simultaneity—almost universally does not feature as a theme in itself.[9]

I will argue that the primary challenge in 'modelling' urban complexity in literary texts lies in the representation of simultaneity, in other words: Strategies of narrating urban complexity are to a considerable extent strategies of narrating simultaneity (for a detailed discussion of simultaneity as central to an understanding of other forms of complexity, cf. Gurr 2011a, b). I will then sketch a number of strategies used in selected literary texts[10] to represent simultaneity.

On the one hand, the representation of co-existing impressions is arguably the crux of any attempt to narrate urban complexity, for simultaneity, the notion of innumerable things—momentous or trivial—happening at the same time, is surely a central characteristic of urban complexity. On the other hand, it is simultaneity which in literary texts poses particular representational challenges. Other key

[8] I discuss this at some length and with numerous examples and references to secondary literature in Gurr 2011a, which is a far more detailed version of this section. Since the present essay appears in the context of a multi-disciplinary engagement with urban complexity , I discuss literary strategies only to the extent necessary to make my point. I therefore also refrain from extensive references to potential further examples.

[9] A few observations on urban complexity are to be found, for instance, in Keunen 2007 and Brandt 2009.

[10] For a recent volume on representations of—general, not necessarily urban—complexity in films, computer games and other media, cf. Eckel et al. 2013. For the city in film, cf. Frahm 2010, 2011, Sanders 2003 and Shiel 2001.

aspects of complexity—multiple causal dependencies, conflicts of interest over the use of space, overlapping and intersecting spatial scales or social and cultural heterogeneity, for instance—pose no principal narrative challenge and lend themselves to being represented by different constellations of characters, for instance. In discussing attempts to represent complexity, from a literary studies perspective, representations of simultaneity are therefore of particular interest.

Lessing's 1766 *Laokoon*, his classic discussion of the "limits of painting and poetry," as the subtitle puts it, is an appropriate entry into a discussion of literary strategies of representing simultaneity. Although Lessing's claims have long been disputed, *Laokoon* is demonstrably still central to discussions of the literary representations of simultaneity.[11] In what is essentially a semiotic argument based on the different sign systems used in painting and literature, Lessing here argues:

> If it is true that painting in its imitation uses means entirely different from those of literature—the former using figures and colours in space, the latter articulated sounds in time—and if the signs unquestionably need to be comfortably related to that which they signify, then signs arranged in contiguity can only designate objects which are contiguous, signs arranged in sequence can only designate objects which appear in sequence or whose parts appear in sequence. (1766, Chap. XVI, pp. 77–78, my translation)

Language, according to this argument, cannot persuasively represent simultaneity "because the coexistence of bodies here collides with the consecutiveness of speech" (XVII, p. 87).[12]

In what is probably the most widely cited passage from the treatise, Lessing summarizes his views on the shortcomings of literature:

> This remains true: The sequence of time is the domain of the poet, just as space is the domain of the painter... If several things which in reality have to be surveyed at once if they are to create a whole were to be narrated to the reader one after the other in order to create a picture of the whole for him, this is an intrusion of the poet into the domain of the painter—and one in which the poet wastes a lot of imagination without any use. (1766, Chap. XVIII, p. 90)

In the course of *Laokoon*, Lessing famously goes on to argue that literature is deficient in representing simultaneity. Thus, we might somewhat provocatively argue that representations of urban complexity must therefore centrally be attempts to prove wrong Lessing's assumption that literature is deficient in the compelling representation of simultaneity.

The following is an attempt—with no claim to completeness—at a sketchy and idealized inventory of strategies of describing, suggesting, or simulating simultaneity.

Arguably the simplest form of representing urban complexity and simultaneity are declarative representations, i.e., passages which state that something is

[11] For discussions of Lessing's *Laokoon* and its implications for semiotics and intermediality, cf. Wellberry 1984 and Koebner 1989.

[12] "weil das Koexistierende des Körpers mit dem Konsekutiven der Rede dabei in Kollision kömmt."

complex without 'showing' or performing complexity. These frequently take the form of 'suggesting' complexity by means of metaphors or similes such as the often clichéd urban 'maze' or 'labyrinth' to express an individual's sense of being lost.[13] Though surely a way of representing the city as 'complex,' this alone is more a thematic notion than a strategy of representation. If the city is merely declared to resemble a maze without representing it as a maze, for our purposes this is not of great interest and as such has little or nothing to do with simultaneity. In the more interesting cases, however, the text will also assume labyrinthine qualities itself or will have the experiential potential of allowing readers to experience the disorientation suggested by the notion of the labyrinth.

A further such strategy is what I term the 'synecdochic representation of simultaneity,' which consists in narrating one strand of action and suggesting that there would have been innumerable others that would also deserve to be told. This technique of highlighting the need to 'select' from countless simultaneous stories equally worth telling is apparent in the much-quoted voice-over conclusion to Jules Dassin's 1948 New York film *The Naked City*: "There are eight million stories in the Naked City. This has been one of them." This phrase is alluded to again with a similar effect in Colson Whitehead's 2003 *The Colossus of New York*: "There are eight million naked cities in this naked city—they dispute and disagree. The New York City you live in is not my New York City; how could it be?... Before you know it, you have your own personal skyline" (p. 6).

What might be called the 'performative or experiential representation of simultaneity' is a central strategy that recurs with remarkable consistency throughout the centuries from eighteenth-century representations of the city all the way to contemporary urban fiction. Complexity and simultaneity are here frequently enacted by means of a suggestive sequence of impressions simulating the chaotic and "rapid crowding of changing images" (Simmel 1903, p. 13), allowing readers to 'experience' the sense of being overpowered by the simultaneity of multiple impressions. One such 'experiential' technique, prevalent in modernist fiction, is the 'stream of consciousness' representation of the overpowering sense impression that frequently occurs in combination with the 'filmic' technique of montage. The reader is here confronted with the attempt at an unfiltered representation of sense perceptions, thoughts, feelings, memories of a character confronted with an urban scene. A passage from the very first pages of Alfred Döblin's *Berlin Alexanderplatz* (1929), which needs to be cited at some length, may serve as a case in point.[14] Franz Biberkopf has just been released after four years in prison and is entering the tram:

---

[13] For a discussion of the city as a maze, labyrinth or jungle, cf. also Brandt 2010, pp. 129–134, 2009, pp. 558–564, Faris 1991, Kelley 1997, Smith 1977, 171 passim, Versluys 2003 as well as several essays in Gurr and Raussert 2011. Cf. also de Certeau's reference to the "mobile and endless labyrinths far below" (1994, p. 92) in the passage on the urban voyeur observing the city from above.

[14] It is interesting to note that even Frank Eckardt begins his reflections on urban complexity with a discussion of a representative passage from Döblin's quintessentially urban novel (cf.

> He shook himself and gulped. He stepped on his own foot. Then, with a run, took a seat in the car. Right among people. Go ahead. At first it was like being at the dentist's, when he has grabbed a root with a pair of forceps, and pulls; the pain grows, your head threatens to burst. He turned his head back towards the red wall, but the car raced on with him along the tracks, and only his head was left in the direction of the prison. The car took a bend; trees and houses intervened. Busy streets emerged, Seestrasse, people got on and off. Something inside him screamed in terror: Look out, look out, it's going to start now. The tip of his nose turned to ice; something was whirring over his cheek. *Zwölf Uhr Mittagszeitung, B. Z., Berliner Illustrierte, Die Funkstunde*. "Anybody else got on?" The coppers have blue uniforms now. … Crowds, what a swarm of people! How they hustle and bustle! My brain needs oiling, it's probably dried up. What was all this? Shoe stores, hat stores, incandescent lamps, saloons. People got to have shoes to run around so much; didn't we have a cobbler's shop out there, let's bear that in mind! Hundreds of polished window-panes, let'em blaze away, are they going to make you afraid or something, why, you can smash'em up, can't you, what's the matter with'em, they're polished clean, that's all. The pavement on Rosenthaler Platz was being torn up; he walked on the wooden planks along with the others. Just go ahead and mix in with people, then everything's going to clear up, and you won't notice anything, you fool. Wax figures stood in the show-windows, in suits, overcoats, with skirts, with shoes and stockings. Outside everything was moving, but—back of it—there was nothing! It—did not—live! It had happy faces, it laughed, waited in twos and threes on the traffic islands opposite Aschinger's, smoked cigarettes, turned the pages of newspapers. Thus it stood there like the street-lamps—and—became more and more rigid. They belonged with the houses, everything white, everything wooden. (1989, pp. 3f.)

In one of the best essays on modernist urban fiction, Bart Keunen discusses *Manhattan Transfer*, which for the present purpose can be regarded as representative of such modernist city novels, and comments on its representation of complexity, once more highlighting the city's "organized complexity" (Jacobs 1961, p. 449):

> The novel shows that Manhattan is a complex whole of overlapping plots (lives) and crossing paths. The different world models converge in a construction that reveals the complexity of the world. It is not a chaotic complexity but a complicated network of individuals, actions, observations, and situations. (Keunen 2001, p. 435)

A further strategy—or rather group of strategies, as the variations are significant—is the attempt to break the linearity of print. This may range from inviting readers to go on reading elsewhere in the book, via recurring phrases in a text which connect different passages in a form of hyperlink avant la lettre (cf. my discussion of Eliot's *The Waste Land* in Gurr 2014, forthcoming), typographical strategies of printing a text in several columns, or the segmentation of the book into unnumbered booklets to be arranged at will, all the way to fully-fledged hypertexts—and more recently to attempts to simulate some of the features and effects of hypertext in print narratives.

(Footnote 14 continued)
Eckardt 2009, 7). However, although it is central to Döblin's endeavor, the notion of simultaneity as a key component of complexity does not play a role at all in Eckardt's study.

One such form of escaping linearity is inherent in the materiality of the book. In recent years, a number of attempts appear to have been inspired by hypertext, but some of the devices are much older, of course. One such technique is to be found in B.S. Johnson's *The Unfortunates* (1969): Its form of presentation—27 chapters ranging from less than a page to 12 pages in length, separately bound and delivered in a box in a random order (only the first and the last chapters are identified as such), so the reader has to choose in what order to read the chapters—appears predominantly designed to escape the linearity of print. Though arguably at least as much an attempt to render emotional complexities and the intricacies of memory, as a novel essentially 'about' Nottingham, this is on some level decidedly also an 'urban' novel.

Arguably one of the formally most inventive, non-linear,[15] and 'complex' fictions in recent decades is Mark Danielewski's *House of Leaves* (2000), a novel centrally concerned with the perception of space and with the literary rendering of architectural features of the built environment. The text cites innumerable architects and urban theorists (there is an eight-page marginal note listing hundreds of architectural monuments (pp. 120–134, even pages), another one listing hundreds of major architects (upside down and backwards from p. 135 to 121, odd pages), a selective bibliography on architecture (152), interspersed philosophical meditations on architecture and the perception of space, numerous references to architecture and urban theory, including, for instance, quotations from Kevin Lynch's *The Image of the City* (p. 176). With its passages printed in mirror writing, countless footnotes, cross-references, parallel columns, typographical games, different colours, musical notations, photographs, sketches, a vast range of intertextual references, appendices, and an (often misleading) index, the novel is a *tour de force* of strategies designed to break linearity, to represent complexity, and to stage simultaneity (though the reader is still forced to choose what to read first).

What these and other strategies and texts share is the attempt to represent various facets of urban complexity from the point of view of literary characters. Yet what may seem aggregable and abstractable (or even negligible entirely) in technical models is precisely what—in contrast to the non-sensual "city of sociologists" (and by implication modelers) Lindner (2008, 92) referred to—makes the city a sensual place that does have its unique sounds, smells, tastes and rhythms.

6 Conclusion

Despite the fundamentally different status of the 'model' in urban complexity modeling on the one hand and in literary and cultural studies on the other hand, there are a number of important parallels and points of intersection between the

[15] Though it is tempting to exploit the notion of 'non-linearity' as a characteristic of both literary and mathematical formulations of urban complexity, I do not currently see how they are related beyond being polysemes.

two types of 'model and in the understanding of complexity: What most technical notions of complexity share is that they measure the complexity of a system in terms of the length or the complexity of the description or representation of that system. A number of complexity theorists have even argued that "complexity is not primarily a characteristic of the object that is being described, but of the description" (Richter and Rost 2002, p. 112; my translation).[16] This notion might lend itself as a bridge between technical or mathematical and cultural conceptualizations of complexity. Ultimately, what is relevant to literary studies, one might argue, is not so much the complexity of the city itself, but the representation of this complexity, i.e., its description in the 'model' of the literary text. Literary studies are thus concerned with the challenge of 'modeling' it, or, in the terminology of literary studies, of 'representing' it. Thus, where technical and mathematical complexity research is concerned with the mathematical description of complexity, literary and cultural studies of urban complexity are concerned with the challenges of verbal representation.

Gell-Mann's notion of "effective complexity" provides a further important connection between a technical and a literary understanding of complexity:

> A measure that corresponds much better to what is usually meant by complexity in ordinary conversation, as well as in scientific discourse, refers not to the length of the most concise description of an entity (which is roughly what AIC is), but to the length of a concise description of a set of the entity's regularities. Thus something almost entirely random, with practically no regularities, would have effective complexity near zero. So would something completely regular, such as a bit string consisting entirely of zeroes. Effective complexity can be high only a region intermediate between total order and complete disorder. (1995a, p. 16)

This seems precisely to be the case with cities: In the sense of "effective complexity", they are systems in which there are multiple connections, but in which by no means everything is directly connected to everyone else and thus systems in which "a concise description of a set of the entity's regularities" would be extremely long. Thus characterized by an intricate combination of both order and disorder, cities have long been understood as systems of extremely high "effective complexity" (take Jane Jacobs's classic formulation that cities are "problems in organized complexity", 1961, p. 449). Somewhat speculatively, we might want to argue that literary texts as models of reality per se are combinations of order and disorder that simulate complexity and multiplicity in frequently highly structured and ordered fashion; they impose order upon disorder and thus structurally replicate key patterns of urban complexity (for a discussion of critical positions on structural analogies between 'city' and 'text', cf. Gurr 2014, forthcoming).

Moreover, if we regard 'scenario building' and the testing of alternative parameter settings in their impact on a given system as a crucial function of urban

---

[16] "Komplexität ist nicht in erster Linie eine Eigenschaft des beschriebenen Objekts, sondern der Beschreibung selbst." (Richter and Rost 2002, 112).

systems modeling, then a further parallel emerges: One of the central functions of literature, according to one understanding, is that it serves as a form of symbolic action, as a social experiment free from the constraints of everyday life—literature as 'depragmatised behaviour in rehearsal' ['entpragmatisiertes Probehandeln'][17] which makes it possible symbolically to try out in fiction different scenarios or potential solutions for key societal issues.[18]

Finally, what is also helpful here are the views of Jürgen Link, Winfried Fluck or Hubert Zapf on functions of literature in the system of culture in the sense of a 'history of its functions' ("Funktionsgeschichte", for an overview cf. Gymnich/ Nünning 2005; for one influential account, cf. Fluck 1997). Thus, following Zapf's suggestive terminology, literature can have the function of a critical cultural diagnosis ("critical meta-discourse"), but it is also an "imaginative counter-discourse," which potentially develops alternatives. Finally, as a "reintegrating inter-discourse" (Zapf 2001, cf. also Zapf 2002), it re-integrates into the cultural whole what is otherwise repressed or marginalized.

What, then, can the study of complex systems learn from literary and cultural studies? It may be the insight that the irreducible element of individual psychological responses to a given urban environment, that human desires, hopes and fears are very often crucial to understanding that environment. What literary and cultural studies as I conceive them can contribute is an understanding of precisely those elements of urban complexity that cannot be measured, modeled, classified or studied in terms of information theory. From the point of view of practical planning, this may mean that what can be planned is very often not that which will make a place distinctive.

My point is not that these elements of the urban should somehow nonetheless be quantified in order to be integrated into the models after all. Rather, I argue that literary texts as an alternative form of 'modeling' urban complexity enable different views and may draw attention to blind spots in other models, thus not only functioning as a type of 'sanity check' but as a further, different and complementary type of 'urban model'.

---

[17] This is the view of formulated, among others, by Kenneth Burke, Dieter Wellershoff, Wolfgang Iser or somewhat more recently in an impressive volume edited by Stefan Horlacher and Stefan Glomb. One classic formulation is Wellershoff's, who spoke of literature as a "space of simulation for alternative behaviour in rehearsal at reduced risk" (Wellershoff 1973, p. 57, my translation): "Simulationsraum für ein alternatives Probehandeln mit herabgesetztem Risiko." Cf. also Glomb and Horlacher 2004, passim. Kenneth Burke's notion of "Literature as Equipment for Living" is a related concept, according to which any work of literature has the social function of being an attempt at naming a situation and coming to terms with it. In this sense, literature can be seen as an assembly of case studies in naming situations and in solving problems, an arsenal of strategies for dealing with situations that is developed in fiction but can lay claim to applicability in life (cf. Burke 1974); for a convenient summary of his position cf. the essay "Literature as Equipment for Living" in that volume.

[18] However, literary texts frequently do not attempt to solve a problem by imposing an answer—and even if they do, they are often less interesting for the answer they propose than for having asked the question and raised the problem.

It is precisely the need to integrate these different perspectives to come to a meaningful understanding of the complex dynamic of 'urban systems' that makes the kind of interdisciplinary dialogue we engage in here so necessary, so challenging and so rewarding.

References

Albeverio, S., Andrei, D., Giordano, P., Cancheri, A. (eds.) The Dynamics of Complex Urban Systems an Interdisciplinary Approach. Physica, Heidelberg (2008)

Assmann, A.: Geschichte findet Stadt. In: Csaky, M., Leitgeb, Ch. (eds.) Kommunikation–Gedächtnis–Raum: Kulturwissenschaften nach dem 'Spatial Turn', pp. 13–27. Transcript, Bielefeld (2009)

Augé, M.: Non-Places: Introduction to an Anthropology of Supermodernity. Verso, London (1995)

Batty, M.: Cities as complex systems: Scaling, interaction, networks, dynamics and urban morphologies. In: Meyers, R.A. (ed.) Encyclopedia of Complexity and Systems Science, pp. 1041–1071. Springer, Berlin (2009)

Berking, H., Löw, M. (eds.): Die Eigenlogik der Städte: Neue Wege für die Stadtforschung. Campus, Frankfurt/Main (2008)

Brandt, S.L.: The city as liminal space: Urban visuality and aesthetic experience in postmodern U.S. literature and cinema. Amerikastudien/Am. Stud. **54**(4), 553–581 (2009)

Brandt, S.L.: Open city, closed space: Metropolitan aesthetics in American literature from Brown to DeLillo. In: Brandt, S.L., Fluck, W., Mehring, F. (eds.) Transcultural Spaces: Challenges of Urbanity, Ecology, and the Environment, pp. 121–144. Gunter Narr, Tübingen (2010)

Burke, K.: The Philosophy of Literary Form: Studies in Symbolic Action. University of California Press, Berkeley (rpt. 1974). Orig. Louisiana State University Press, Baton Rouge (1941)

Butler, M., Gurr, J.M.: Urbane Populärkultur als Bewertungspraxis und -ressource: Zum normativen Potential populärkultureller Inszenierung und zur diskursiven Aneignung urbaner Räume. In: Warnke, I., Busse, B. (eds.) Bewertung urbaner Räume. Akademie Verlag, Berlin (2014, forthcoming)

Byrne, D.: Complexity Theory and the Social Sciences: An Introduction. Routledge, London (1998)

Cohen, J., Stewart, I.: The Collapse of Chaos: Discovering Simplicity in a Complex World. Penguin, Harmondsworth (1995)

Danielewski, M.Z.: House of Leaves. Pantheon, New York (2000)

de Certeau, M.: The Practice of Everyday Life (Trans: Steven Rendall). University of California Press, Berkeley (1994)

Dillon, S.: The Palimpsest: Literature, Criticism, Theory. Continuum, New York (2007)

Döblin, A.: Berlin Alexanderplatz: The Story of Franz Biberkopf (Trans: Eugene Jolas). Continuum, New York (1989)

Eckardt, F.: Die komplexe Stadt: Orientierungen im urbanen Labyrinth. Verlag für Sozialwissenschaften, Wiesbaden (2009)

Eckel, J., Leiendecker, B., Olek, D., Piepiorka, Ch. (eds.): (Dis)Orienting Media and Narrative Mazes. Transcipt, Bielefeld (2013)

Faris, W.B.: The labyrinth as sign of city, text, and thought. In: Caws, M.A. (ed.) City Images: Perspectives from Literature, Philosophy, and Film, pp. 32–41. Gordon and Breach, New York (1991)

Fluck, W.: Das kulturelle Imaginäre: Eine Funktionsgeschichte des amerikanischen Romans, 1790-1900. Suhrkamp, Frankfurt/M (1997)

Frahm, L.: Jenseits des Raums: Zur filmischen Topologie des Urbanen. Transcript, Bielefeld (2010)
Frahm, L.: Reading the fragments: Toward an urban topology of episodic films. In: Hassenpflug, D., Giersig, N., Stratmann, B. (eds.) Reading the City: Developing Urban Hermeneutics/Stadt lesen: Beiträge zu einer urbanen Hermeneutik, pp. 207–226. Verlag der Bauhaus-Universität, Weimar (2011)
Gell-Mann, M.: What is complexity? Complexity **1**(1), 16–19 (1995)
Gell-Mann, M.: The Quark and the Jaguar: Adventures in the Simple and the Complex. Abacus, London (1995b)
Glomb, S., Horlacher, S. (eds.): Beyond Extremes: Repräsentation und Reflexion von Modernisierungsprozessen im zeitgenössischen britischen Roman. Gunter Narr, Tübingen (2004)
Gurr, J.M.: The literary representation of urban complexity and the problem of simultaneity: a sketchy inventory of strategies. In: Gurr, J.M., Raussert, W. (eds.) Cityscapes in the Americas and Beyond, pp. 11–36. WVT and Bilingual Press, Trier and Tempe (2011a)
Gurr, J.M.: The politics of representation in hypertext docufiction: Multi-ethnic Los Angeles as an emblem of 'America' in Norman M. Klein's Bleeding Through: Layers of Los Angeles 1920–1986. In: Raab, J., Thies, S., Noll-Opitz, J. (eds.) Screening the Americas: Narration of Nation in Documentary Film, pp. 153–171. WVT and Bilingual Press, Trier and Tempe (2011b)
Gurr, J.M.: The modernist poetics of urban memory and the structural analogies between 'city' and 'text': The Waste Land and Benjamin's Arcades Project. In: Freitag, K. (ed.). Recovery and Transgression: Memory in American Poetry. Cambridge Scholars Publishing, Newcastle (2014, forthcoming)
Gurr, J.M., Butler, M.: Against the 'erasure of memory' in Los Angeles city planning: Strategies of re-ethnicizing L.A. in digital fiction. In: Kaltmeier, O. (ed.) Selling EthniCity: Urban Cultural Politics in the Americas, pp. 145–163. Ashgate, London (2011)
Gurr, J.M., Butler. M.: On the 'cultural dimension of sustainability' in urban systems: Urban cultures as ecological 'force-fields' in processes of sustainable development. In: Neis, H., Brown, G., Gurr, J.M., Schmidt, J.A. (eds.) Generative Process, Patterns, and the Urban Challenge: Fall 2011 International PUARL Conference, pp. 77–86. PUARL Press, Portland (2012)
Gurr, J.M., Raussert, W. (eds.): Cityscapes in the Americas and Beyond: Representations of Urban Complexity in Literature and Film. WVT and Bilingual Press, Trier and Tempe (2011)
Gymnich, M., Nünning, A. (eds.): Funktionen von Literatur: Theoretische Grundlagen und Modellinterpretationen. WVT, Trier (2005)
Haken, H., Portugali, J.: The face of the city is its information. J. Environ. Psychol. **23**, 385–408 (2003)
Harvey, D.: The Condition of Postmodernity: An Inquiry into the Origins of Cultural Change. Blackwell, London (1989)
Hassenpflug, D.: Once again: Can urban space be read? In: Hassenpflug, D., Giersig, N., Stratmann, B. (eds.) Reading the City: Developing Urban Hermeneutics/Stadt lesen: Beiträge zu einer urbanen Hermeneutik, pp. 49–58. Verlag der Bauhaus-Universität, Weimar (2011)
Hassenpflug, D.: Walter Benjamin und die Traumseite der Stadt. In: Hassenpflug, D. Reflexive Urbanistik: Reden und Aufsätze zur europäischen Stadt, pp. 7–22. Verlag der Bauhaus-Universität, Weimar (2006)
Huyssen, A.: Present Pasts: Urban Palimpsests and the Politics of Memory. Stanford University Press, Palo Alto (2003)
Jacobs, J.: The Death and Life of Great American Cities. Random House, New York (1961)
Johnson, B.S.: The Unfortunates. New Directions, New York (2008). 1969
Kelley, W.: Pierre in a labyrinth: The mysteries and miseries of New York. In: Bryant, J., Milder, R. (eds.) Melville's Ever Moving Dawn: Centennial Essays, pp. 393–405. Kent State University Press, Kent (1997)

Keunen, B.: The plurality of chronotopes in the modernist city novel: The case of Manhattan Transfer. Engl. Stud. **82**(5), 420–436 (2001)

Keunen, B.: Living with fragments: World making in modernist city literature. In: Eysteinsson, A., Liska, V. (eds.) Modernism, pp. 271–290. John Benjamins, Amsterdam (2007)

Koebner, Th (ed.): Laokoon und kein Ende: Der Wettstreit der Künste. Edition Text+Kritik, Munich (1989)

Lefebvre, H.: The Production of Space (Trans: Donald Nicholson-Smith). Blackwell, Oxford (1991)

Lessing, G.E.: Laokoon oder über die Grenzen der Malerei und Poesie mit beiläufigen Erläuterungen verschiedener Punkte der alten Kunstgeschichte [1766]. Lessing. Ausgewählte Werke. Ed. W. Stammler, vol. 3, pp. 1–150. Carl Hanser, Munich (1959) [1766]

Li, M., Vitányi, P.: An Introduction to Kolmogorov Complexity and its Applications. Springer, New York (1993)

Lindner, R.: Textur, imaginaire, Habitus—Schlüsselbegriffe der kulturanalytischen Stadtforschung. In: Berking, H., Löw, M. (eds.) Eigenlogik der Städte: Neue Wege für die Stadtforschung, pp. 83–94. Campus, Frankfurt/M. (2008)

Lynch, K.: The Image of the City. MIT Press, Cambridge (1960)

Mainzer, K.: Thinking in Complexity: The Computational Dynamics of Matter, Mind and Mankind. Springer, Berlin (2007)

Martindale, Ch.: Ruins of Rome: T.S. Eliot and the presence of the past. Arion **3**(2–3), 102–140 (1995)

Portugali, J.: Self-Organization and the City. Springer, Berlin/Heidelberg (2000)

Portugali, J.: Complexity theory as a link between space and place. Environ. Plan. A **38**, 647–664 (2006)

Portugali, J.: Complexity, Cognition and the City. Springer, Berlin/Heidelberg (2011)

Portugali, J.: Complexity theories of cities: Achievements, criticism and potentials. In: Portugali, J., Meyer, H., Stolk, E., Tan, E. (eds.) Complexity Theories of Cities Have Come of Age: An Overview with Implications to Urban Planning and Design, pp. 47–62. Springer, Berlin/Heidelberg (2012)

Read, S.: Meaning and material: Phenomenology, complexity, science and 'adjacent possibilities'. In: Portugali, J., Meyer, H., Stolk, E., Tan, E. (eds.) Complexity Theories of Cities Have Come of Age: An Overview with Implications to Urban Planning and Design, pp. 105–127. Springer, Berlin/Heidelberg (2012)

Richter, K., Rost, J.-M.: Komplexe Systeme. Fischer, Frankfurt (2002)

Sanders, J.: Celluloid Skyline: New York and the Movies. Alfred A. Knopf, New York (2003)

Sharpe, W., Wallock, L.: From 'great town' to 'nonplace urban realm': Reading the modern city. In: Sharpe, W., Wallock, L. (eds.) Visions of the Modern City: Essays in History, Art, and Literature, pp. 1–50. Johns Hopkins University Press, Baltimore (1987)

Shiel, M.: Cinema and the city in history and theory. In: Shiel, M., Fitzmaurice, T. (eds.) Cinema and the City: Film and Urban Societies in a Global Context, pp. 1–19. Blackwell, Malden (2001)

Simmel, G.: The metropolis and mental life. In: Miles, M., Hall, T., Borden, I. (eds.) The City Cultures Reader. Routledge, London, pp. 12–19 (2004) [1903]

Smith, P.F.: The Syntax of Cities. Hutchinson, London (1977)

Suttles, D.: The cumulative texture of local urban life. Am. J. Sociol. **90**(2), 283–304 (1984)

Versluys, K.: New York as a maze: Siri Hustvedt's The Blindfold. In: Lenz, G., Riese, U. (eds.) Postmodern New York City: Transfiguring Spaces—Raum-Transformationen, pp. 99–108. Winter, Heidelberg (2003)

Wellbery, D.E.: Lessing's Laocoon: Semiotics and Aesthetics in the Age of Reason. Cambridge University Press, Cambridge (1984)

Wellershoff, D.: Literatur und Lustprinzip. Kiepenheuer & Witsch, Cologne (1973)

Whitehead, C.: The Colossus of New York: A City in Thirteen Parts. Doubleday, New York (2003)
Zapf, H.: Literatur als kulturelle Ökologie: Zur kulturellen Funktion imaginativer Texte an Beispielen des amerikanischen Romans. Niemeyer, Tübingen (2002)
Zapf, H.: Literature as cultural ecology: notes towards a functional theory of imaginative texts with examples from American literature. In: Grabes, H. (ed.) Literary History/Cultural History: Force-Fields and Tensions. REAL 17 (Research in English and American Literature), pp. 85–99. Narr, Tübingen (2001)

Film

The Naked City (Dir.) Jules Dassin. U.S.A. (1948)

About the Authors

Funda Atun Funda Atun, born in 1983, PhD in Spatial Planning and Urban Development. She is currently a post-doc researcher at Politecnico di Milano in a European Union FP7 project, named "KNOW 4 DRR", in which the team elaborates a knowledge management system specialized in disaster risk research. Her research interests include disaster risk management, earthquake risk assessment, flood risk assessment, land use planning, and complex systems approaches.

Ernst Gebetsroither-Geringer Ernst Gebetsroither-Geringer, born 1970, studied Ecology and has been scientist at the AIT Austrian Institute of Technology since January 1999. His main research interests include modeling and simulation of complex systems with a special focus on spatial modeling and combining different modeling methods. For several years, he has taught the Universities of Vienna and Graz. Recently, he started to focus on urban energy management and the development of urban planning support tools in the context of Smart Cites. He is also an active member within the European Energy Research Alliance (EERA) in the context of an urban simulation platform development.

Jens Martin Gurr Jens Martin Gurr, born 1974, has been Chair of British and Anglophone Literature and Culture at the University of Duisburg-Essen since April 2007. His research interests include urban literary and cultural studies, interdisciplinary urban systems research, contemporary Anglophone literature, seventeenth and eighteenth century British literature and British Romanticism. Since 2008, he has been director of the University of Duisburg-Essen's Joint Center "Urban Systems"; he is also director of the international interdisciplinary doctoral programme "ARUS–Advanced Research in Urban Systems". He is the author of two monographs and about 60 essays and articles as well as co-editor of eight collections of essays.

Tom Kauko Tom Kauko's research interest is in pricing and sustainability aspects of urban real estate. He is the leader of the Urban, Rural and Regional research group at the Department of Geography, Norwegian University of Science

C. Walloth et al. (eds.), *Understanding Complex Urban Systems: Multidisciplinary Approaches to Modeling*, Understanding Complex Systems,
DOI: 10.1007/978-3-319-02996-2, © Springer International Publishing Switzerland 2014

and Technology (NTNU). He has been Associate Professor in Urban Geography at NTNU since October 2006. He received a M.Sc. degree in Real Estate in 1994 (Helsinki University of Tech, Finland) and a Ph.D. in Geography in 2002 (Utrecht University, The Netherlands). He has worked in research for several universities in four different countries for more than fifteen years. His main fields of research are property value, mainly concerning the residential sector, and sustainable built environment, mainly urban regeneration. He has over 40 articles in international peer-reviewed journals plus a number of monographs and book chapters.

Xosé Lois Martínez Suárez Xosé Lois Martínez Suárez is Professor of Urban Planning at the University of A Coruña (UDC), Spain. His more than 70 publications include eight books on the issues of architecture, urban planning and urban history. He has been Director on the Editorial Board of the journal Revista de Arquitectura OBRADOIRO, (Official College of Architects of Galicia-COAG) (1990–2000) and Chairman of the Committee of Culture of COAG-A Coruña. He has served on the Historical Heritage Commission of Galicia (1985–1989 and 2003–2010). As Mayor-City Council President of Oleiros (1979–1983), he promoted its first General Urban Plan.

José Pérez Ríos José Pérez Ríos is Professor of Business Organization at the University of Valladolid (UVA), Spain. His research focuses on the application of management cybernetics and system dynamics to the study of complex systems. He created the "Navegador Colón" for the Spanish Foreign Ministry, and the VSMod® software (VSM application). He has worked in multiple research projects and written more than 80 publications. His five books to date include *Design and Diagnosis for Sustainable Organizations. The Viable System Method* (Springer, 2012). He is member of the Board of Directors of the WOSC (World Organisation of Systems and Cybernetics).

Rüdiger Schultz Rüdiger Schultz, born 1959, has been Full Professor for Discrete Mathematics and Optimization at the Gerhard-Mercator University Duisburg, today University of Duisburg-Essen, since September 1998. His main fields of research are Stochastic Optimization, Mixed-Integer Optimization, and Industrial Applications of Mathematical Optimization; the latter mainly in the power and gas industries and, more recently, in logistics. Schultz is the author or co-author of about 60 peer-reviewed research articles. He serves on the editorial boards of seven scientific journals in mathematical optimization and operations research.

Christian Walloth Christian Walloth, born 1979, has been a PhD student in the international interdisciplinary doctoral programme "ARUS–Advanced Research in Urban Systems" since 2011. His research interests include complex urban systems, in particular the emergence of real novelty (new qualities) in complex systems, and the possibilities of dealing with emergent systems, specifically possibilities of applying methods of effectuation to urban development. Christian Walloth is chair of the bi-annual symposium of "Urban Systems Research" at the "EMCSR–European Meetings on Cybernetics and Systems Research" and editor of the book series on "Interdisciplinary Approaches to Modeling Complex Urban Systems".

Hans-Werner Wehling Hans-Werner Wehling, born 1949, holds the professorship of regional geography of old-industrial regions. His research interests include the regional geography and planning of the Ruhr region and of Britain as well as studies in historical and current urban geography. He is a member of the University of Duisburg-Essen's Joint Center "Urban Systems" and a Fellow of the Royal Geographical Society, London

Index

C. Walloth et al. (eds.), *Understanding Complex Urban Systems: Multidisciplinary Approaches to Modeling*, Understanding Complex Systems,
DOI: 10.1007/978-3-319-02996-2, © Springer International Publishing Switzerland 2014

MIX
Papier aus verantwortungsvollen Quellen
Paper from responsible sources
FSC® C105338

Printed by Books on Demand, Germany